The
Everyday
Naturalist

This book was written on unceded Chinook land on the Long Beach Peninsula in southwest Washington. The Chinook Indian Nation has been fighting for more than half a century for federal recognition. While it was briefly recognized in 2001, eighteen months later, its recognition was rescinded. To learn how you can help the Chinook fight to regain recognition and thereby gain access to much-needed resources, please visit ChinookJustice.org.

The *Everyday* Naturalist

How to Identify Animals, Plants, and Fungi Wherever You Go

REBECCA LEXA

Illustrations by Ricardo Macía Lalinde

TEN SPEED PRESS
California | New York

CONTENTS

INTRODUCTION

A Toolkit for Nature Nerds

sage thrasher (*Oreoscoptes montanus*)

ONE OF MY BEST MOMENTS EVER WAS WHEN I GOT TO add a bird to my life list and impress a guy at the same time.

Okay, so the guy was my longtime partner, who's impressed by everything I do, and he once told me, "I need a life list of animals I've looked up on Google Image Search because you told me you saw one in the wild." But I really did outdo myself this time. We were out on a chilly March day near Echo, Oregon, tracing the ruts left by wagons on the Oregon Trail. Small flushes of greenery were just starting to emerge beneath miles of twisted, pungent sagebrush scrub, and we stopped every so often to watch large black darkling beetles trundle across our path. They—and one pile of dried-up coyote scat—looked to be our only wildlife sightings for the day.

That was, until a flash of feathers snapped my attention to the northwest. Forty yards off and flying away fast was a towhee-sized bird; even at that distance, I could see its grayish-brown back and the white belly flecked with an abundance of brown spots. But it was the bill that I noticed the most: slender and longer than that of a thrush, with a slight curve. It was not so large as the one on the brown thrasher I had seen near my parents' place in Missouri the year before, and it lacked that bird's russet tones. The overall giss, though, indicated "thrasher."*

The bird quickly winged its way low over the scrub before it disappeared into the sea of sage. All told, I had it in view for about three seconds. But it was enough to trigger memories of pictures I'd seen while paging through countless field guides and articles. In this area, at this time of year, there was only one bird it could be.

"Sage thrasher!" I yelled to my partner, pointing where it had gone. "My first one!"

He was momentarily incredulous. "You're kidding. You've never seen one in person before?"

"I am absolutely certain of it." I giddily recited the traits that pointed toward my identification.

He shook his head and grinned. After more than a decade together, he was used to me getting distracted every five feet so I could pin an identification on some animal, plant, or fungus I'd seen. But this was proof I'd leveled up: I'd nailed down my unidentified flying object with mere seconds of observation time, at a distance, and with no prior in-person experience with this species.

I did, of course, verify my identification once we got back to the car, pulling out field guides and checking websites of the National Audubon Society, the Cornell Lab of Ornithology, and other well-respected sources. I checked iNaturalist to see if anyone else had seen sage thrashers here this early in the year, since they are in the Pacific Northwest only for the breeding season. I looked up similar species that might be found in this location at this time, and none were as close a match with the bird I saw. Everything still pointed to my encounter as seeing one of the earliest sage thrashers to return to Oregon for the breeding season. I'm always

---

* *Giss* (sometimes given the more unfortunate spelling "jizz") is derived from *gestalt*, and it is a term used by birders to describe the overall appearance or impression of a bird. It's often more qualitative than quantitative, although it's informed by individual traits that all add up to what the viewer thinks they've seen. Sometimes you can get a detailed description, but often the giss boils down to "It just had that warbler look to it, you know?"

open to the possibility that I am wrong no matter how solid my research is, but for now, I'm certain enough that I added the sage thrasher to my life list, thanks to that encounter.

Being able to identify a bird at fifty paces may seem like a small accomplishment, but for me it was a major milestone on a dedicated path that I've been exploring for the better part of two decades. I've always been a giant nature nerd ever since I was big enough to toddle around the yard. But when I left behind the Ozark oak-hickory forests in my twenties and landed in the coniferous Pacific Northwest, I realized I was in unfamiliar territory.

I countered my disorientation by learning the unknown species around me. At first, it was just a few each time I went hiking, but as the years went on, my quest intensified. I purchased stacks of field guides and scoured websites, and when I was introduced to iNaturalist as part of a citizen science project, I added electronic apps to my toolkit.

And now I'm handing that toolkit over to you. Whether you are a seasoned naturalist or just starting to learn the species around you, I want you to feel more confident in your ability to positively identify a new-to-you species. You aren't required to have a degree in the natural sciences to do this, either. *The Everyday Naturalist* is full of techniques and tools that anyone can use, both to learn about your neighborhood nature and when traveling to exciting new ecosystems.

In an age where there's an app for (almost) everything and AI seems poised to replace everyone, it may be tempting to just download PictureThis or Google Lens and be done with it. But these are only tools and imperfect ones at that. If you're going to be really good at nature identification, you need to also develop the critical thinking skills necessary to determine whether the tool you're using is giving you an accurate ID or not.

That's why this book takes my entire identification process and breaks it down into components that are easy to understand and use. I start by explaining the benefits of being able to identify the nature around you, and why it's important to take your time and be thorough. A chapter on the Western naturalist tradition (as one of many naturalist traditions throughout history and around the world) offers more context for why we categorize and learn about living beings the way we do. Then we get into the nuts and bolts of what you'll need to get started with identification and why these tools are useful, from field guides and apps full of species to peruse to other practical tools like binoculars, hand lenses, and the best vessels to carry specimens home for study.

After that, it's time to discuss what you need to pay attention to when you're identifying an unknown organism. Individual chapters about identifying animals, plants, and fungi go into more detail about what sets each of these kingdoms apart from one another. A few organisms, like lichens and slime molds, don't fit into these categories so easily, so they get a chapter all their own. Detailed case studies based on my own experiences with various species demonstrate how to use all these traits to identify sample species across multiple kingdoms, and Recommended Resources on page 257 offer suggestions for further reading and resources. A glossary on page 252 defines some of the specialized terms I've used throughout the book. To view the various books, websites, and sources I used when writing this book, check out my bibliography at rebeccalexa.com/bibliography. And finally, I've created some templates that you can copy and take into the field with you (see pages 246–251). These act as a checklist for all the traits I talk about in the book and make taking notes easier, too.

While I wrote this book to be loosely based on the structure of my two-day *Nature Identification for the Everyday Naturalist* class, you're welcome to read it in whatever order you like. If the first two chapters on theory and context just don't interest you, for example, feel free to jump ahead to the identification chapters. Or let's say you bought this mainly to help you improve your mushroom identification chops—no one's stopping you from starting with the chapter on fungi. I do recommend eventually getting around to reading the whole book, even if it isn't in order.

Just like any other book, this one is also only a tool, and its effectiveness is going to depend on how carefully you use the material within to improve your observation and critical thinking skills. My advice for you is this: Be patient. Don't rush things. Your goal should be as solid an identification as possible, whether that is down to the species level or not, and that can take time. The more you practice nature identification, the better you'll get, and every organism you meet is a chance to put your skills to work.

So, let's get started!

CHAPTER 1

Why Do We Need Everyday Naturalists?

coast redwood trees (*Sequoia sempervirens*)

RAISE YOUR HAND IF, LIKE ME, YOUR ABYSMAL LACK of math skills prevented you from getting a degree in the natural sciences, and you are envious of your friends who get to be field biologists, restoration ecologists, and other professional scientists.

Okay, thank you. I don't feel so alone now. And I have good news for all of us who missed out on our chance to focus on biology, ecology, and related fields in our educational paths: You don't have to have a science, technology, engineering, and math (STEM) degree—or any degree, in fact—to be an everyday naturalist.

A naturalist studies natural history, which is based primarily on observation rather than experiments. Naturalists sometimes focus on a given species or habitat, but they do not separate it from the whole environment. Whereas modern science has increasingly divided itself into ever more specialized areas of focus, natural history brings everything back together into a big-picture view. It encourages long-term, in-depth study of the beings within their ecosystems with a complex understanding of each of their life cycles and stages, behaviors, interactions with other species, and so on.

This is similar to what an ecologist does, although that title is generally reserved for someone who has earned academic degrees in some form of ecology. While ecology is a multidisciplinary field that draws on many different areas of science to understand an entire ecosystem, specialization is becoming increasingly common, such as studying a single species, or an application like restoration ecology. The roots of ecology are within natural history, but forces ranging from scientific trends to a lack of funding are causing ecologists to increasingly abandon long-term studies that give a comprehensive look at a given species or ecosystem.

Another thing that sets natural history apart is its relative egalitarianism. Professional scientists are almost entirely people with (usually multiple) STEM degrees and internships under their belts, and competition for even entry-level jobs can be stiff. Naturalists, on the other hand, come from a much wider variety of professional and experiential backgrounds. One naturalist may be a professional, degreed wildlife biologist with twenty years of experience in the field, while another never went to college but has spent the same twenty-year period painstakingly studying the "herps"—reptiles and amphibians—of their region.

You probably won't see the latter person in charge of a formal scientific study on amphibian population responses to environmental pollutants, simply because no institution would give a person with zero professional credentials the necessary funding. But it's likely that they have at least as deep an understanding of their local herp species as the professional biologist. In fact, find any group of herp enthusiasts, mushroom hunters, or native plant advocates, and you're likely to run into plenty of "amateurs" who could give the academics a run for their money.

I use the term *everyday naturalist* to refer to folks like these. It's similar to the term *citizen scientist,* a member of the general public who volunteers time helping professional scientists in their research.* But whereas citizen science is usually performed in

---

* There is recent debate as to whether "community science" is a more accurate and inclusive term. While I agree with the arguments for the change, I have chosen to use citizen science terminology because it is more widely recognizable and some of the resources later in the book use the term, and I'd like to avoid confusion.

conjunction with specific scientific studies and surveys, the everyday naturalist is constantly observing and learning about the nature around them, regardless of experience, education, or background. A pack of everyday naturalists on the loose might include retirees curious about birdwatching, field biologists on their day off, seasoned mushroom foragers, and that one person who knows every cool bug they encounter.

The Importance of Nature Identification

If you don't know what organisms you're observing, it's going to be a lot harder to compile a picture of what's happening in a given ecosystem. The everyday naturalist is a generalist, and every organism, niche, and ecosystem may be of interest. You might have certain species you particularly like or specific places that especially fascinate you, and there's nothing wrong with giving these extra attention. But ultimately the goal of this book is to prepare you—as the subtitle states—to identify animals, plants, and fungi wherever you go.

You're going to hear me say this again and again: The more you practice identification, the better you get at it. If you stick just to familiar places or taxa, you can challenge yourself only so far. By getting out of your comfort zone, you're sharpening those identification skills, which makes it more likely that your future identifications will be more accurate. You'll also develop a better understanding of both the process of identification and how to use the tools available to you. With time, it may take you only a few minutes to examine a new organism, look it up in a field guide, and then verify it with other sources.

You may also get to encounter entirely new types of organisms you didn't even know existed. I lead guided nature tours in the Pacific Northwest, and on multiple occasions, I have had the pleasure of introducing people to their first slime mold (and explaining that, no, it's not a fungus). The sheer delight these folks have upon realizing that this is something entirely new to them never gets old. That experience is a big part of what motivates me to visit new ecosystems whenever I can. But even if I stick close to home, there's no telling what surprises might be waiting for me in places I've been a thousand times.

Positively identifying a species means you can learn more about it and how it fits into its ecosystem, exploring its relationships with other species, how it uses its habitat, and even whether it is endangered. And if an unknown species closely resembles a familiar one, you may be able to figure out what you've found that much more quickly. The first time I ever saw a varied thrush (*Ixoreus naevius*) with its bold orange-and-black-patterned plumage, I thought, "Wow, that looks a lot like a robin, but more punk rock!" Since the American robin (*Turdus migratorius*) is a member of the thrush family, I opened my field guide to that section. Sure enough, there was my mystery bird!

And, perhaps most importantly, you're expanding your knowledge about and experience of the world, which is beneficial on a multitude of levels. You gain a greater understanding of how the ecosystems you explore are connected and what makes them unique. A lifetime spent learning can be personally fulfilling, and it gives you goals to pursue no matter your age or interests. It's also great for your brain's health and can lower your risk of memory decline and other cognitive issues as you age.

For me, nature identification is a part of how I constantly engage with my surroundings. For example, I think of where I live not just in terms of my street address, but also how I am at the junction of a patch of mature forest, a pasture full of non-native grasses, and a freshwater lake surrounded by wetlands and riparian communities. If I walk around the farm I live on, my awareness of where I am on the land is based not just on buildings and fences, but also where the coastal dunes transition to early-succession shore pine forest and then to older mixed conifers. I am more likely to think of a spot on the farm as "where purple spindles (*Alloclavaria purpurea*) first show up in autumn" than "just south of the beach house's sunroom."

On long road trips, I mark my progress not only with mile markers, but also by the landscapes along the highway and their plant communities, along with any wild animals I might see. Put me someplace new, and I'm going to start looking for the most common species I find there and try to learn them as indicators of this new place. Knowing the nature around me makes me feel more comfortable, and I've never met an ecosystem that I didn't fall in love with once we were familiar with each other.

Similar to biogeography, which defines areas of land like ecoregions according to what beings live there, nature identification offers a more ecologically based understanding of where you are in the world. We are still a part of the community of nature, even in the middle of a city, and identification helps you to know who your nonhuman neighbors are. This knowledge is just as important for them as it is for you.

How Does Identification Help Nature?

Reading the news as a nature lover can be a pretty grim experience. It seems like every week there's some major environmental disaster, be it an oil spill, wildfire, or loss of yet another iconic species. To find these stories, I usually have to sift through several others about the growing impacts of climate change, like the thawing of Arctic permafrost or increasingly chaotic weather patterns. I understand how people often become numb to the headlines out of sheer self-defense because all that bad news can literally be traumatizing.

One symptom of that trauma that I've seen is the inability to put all that bad news aside to just enjoy being in nature. I'll walk outside on a beautiful sunny day, and all I can think about is how climate change made that day warmer and drier than usual. Or I'm hiking along a wooded trail, and the fact that all the trees are less than fifty years old is a stark reminder that this was once an old-growth forest that was clear-cut. And don't even get me started on how, once I recognize an invasive species, I can't stop seeing it *everywhere*.

But I think it's important for us to consciously work toward compartmentalizing our various experiences of nature. We need to be able to set aside the anxiety and grief over environmental destruction in part for our own mental health. Too much of that leads to burnout, which is no good for someone who wants to keep working toward a better world. More importantly, time in nature should be restorative, and if we're too preoccupied with worry, we aren't going to be as receptive to the positive benefits nature has—both physically and psychologically.

Nature identification is one way to give us a better focus when we're outdoors, a necessary distraction from the doom and gloom. This shift in focus to something more positive and constructive can lead to increased resilience and improved mental well-being, giving our bodies and brains a break from the stress of being too aware of current events. Additionally, we may renew our commitment to protecting the species we identify and the habitat they rely on because they are so much more present in our everyday lives.

Knowing an organism's name makes it more personal. Take plants, for example. Many people can identify at least a few animal species but see the plants around them as a single green curtain rather than a community of diverse living beings. This phenomenon is known as plant awareness disparity, and it leads to the frequent view that plants are just the setting in which the dramas of animals play out.* But once you start to learn the names of individual plant species, they stand out from the crowd, and it becomes increasingly apparent that we are immersed in a veritable symphony of Plantae.

That shift in awareness doesn't necessarily stop with one's own appreciation of the biodiversity found in a forest or field. When something is personal, it becomes more important to you. A meadow you've driven by a thousand times and never thought

---

* You might be more familiar with the term *plant blindness*, which is falling out of favor as it suggests there is something wrong with being blind. Biologist Kathryn M. Parsley, who is herself visually impaired, suggests *plant awareness disparity* is a better alternative term.

of twice may become that place where you can walk amid Queen Anne's lace (*Daucus carota*) and chicory (*Cichorium intybus*) growing side by side with tufted hairgrass (*Deschampsia cespitosa*), while black-capped chickadees (*Poecile atricapillus*) forage under nearby shrubs.

And we protect what's important to us. Perhaps you'll be part of a local movement to protect that meadow as a park for all to enjoy, rather than yet another strip mall or subdivision. If successful, you could get the opportunity to help plant more native plants there as part of a habitat restoration. Get to know enough species and your expertise might lead you to volunteering as a docent, introducing visitors to those same amazing species you've discovered there over time.

But this sense of responsibility works on larger scales, too. The identification of key species, including those that are rare or have particularly important ecological roles, can be crucial to making sure an ecosystem is protected. Encouraging the general public to notice endangered species and report them helps get people engaged with conservation in a personal, everyday way. The same goes for having folks be on the lookout for invasive species, too—especially those that are in danger of spreading to new areas.

The devastation of nature worldwide has happened precisely because we have allowed it to become so impersonal that most of us don't think twice about the damage being done. Those of us who do care often find ourselves needing to emotionally distance ourselves just to survive. Because familiarity is antithetical to callous and wanton destruction, nature identification can be a tool for conservation. As more people reengage with nature by learning who their nonhuman neighbors are, they reclaim responsibility for that entire community. It is my hope that this will help lead us toward a more balanced and sustainable relationship between humanity and the rest of nature—starting with knowing their names.

CHAPTER 2

*A Brief History of Nature Identification in Western Science**

Colombian white-faced capuchin monkeys (*Cebus capucinus*)

THERE ARE MANY THINGS I LOVE ABOUT HUMANITY: our creativity, our empathy with not just each other but other species as well, and our curiosity about the world. We're not without our biases and errors, of course, but I appreciate that so many people embody the best of what we have to offer. One of my favorite things about us is that when we find something really nifty, we want to immediately go and share it with other people. We're not just showing off our find, we're expanding our excitement beyond ourselves. And then once we've done that, we examine it to learn more about it.

Our ancient primate ancestors may have been concerned mainly with problems like "Is this edible?" or "Will it hurt me?" or even "Is it shiny enough to help me attract a mate?" These are

_____

* I owe a great deal to David Bainbridge's *How Zoologists Organize Things: The Art of Classification* (Frances Lincoln, 2020) for the inspiration for and organization of this chapter. If you enjoy my writing here and want to find out more—including hundreds of images of natural history illustration throughout Western history—I recommend picking up the aforementioned book, as well as his *Stripped Bare: The Art of Animal Anatomy* (Princeton University Press, 2018).

certainly still important questions for modern humans, but we've expanded our categorization of all the parts of our world far beyond basic survival. That includes organizing every known life-form in ways that improve our understanding of them.

KINGDOM
PHYLUM
CLASS
ORDER
FAMILY
GENUS
SPECIES

Today, we use a system known as taxonomy. Broadly speaking, taxonomy is a way of organizing things into related groups, often by nesting them into a hierarchy that adds dimension and detail. Within biology, taxonomy primarily concerns how various living beings may be organized based on how they are related, a subsection of phylogenetics. From largest to smallest, here are the seven taxonomic ranks you are most likely to encounter in your identification efforts: kingdom, phylum, class, order, family, genus, and species. Others you might commonly run into are domain (one level higher than kingdom), superorder (tucked between class and order), subspecies (a geographically distinct subsection within a species), clade (a group of beings all sharing a common ancestor), and tribe (sometimes used between family and genus). There are several other subspecific divisions, such as variety, form, and ecotype; all these simply refer to populations within a species that have unique characteristics but that can still interbreed with other members of that species.

There are several mnemonics people have created to help remember the seven common taxonomic ranks. Here is the one I first learned:

King (kingdom)

Phillip (phylum)

Can (class)

Order (order)

Five (family)

Great (genus)

Species (species)

Other fun ones I've seen include Keeping Precious Creatures Organized For Grumpy Scientists, Kathryn Proudly Called One Faithful German Shepherd, and Krabby Patties Cook On Fry Grills, Spongebob.

Here's how to classify our own species, humans, using the seven common ranks: Kingdom: Animalia (animals), Phylum: Chordata (animals with spinal cords), Class: Mammalia (mammals), Order: Primates (apes, monkeys, lemurs, lorises, and other primates), Family: Hominidae (great apes), Genus: *Homo* (humans, including extinct species like *H. neanderthalensis*), Species: *H. sapiens* (that's us!).

An organism's scientific name (also known as its binomial nomenclature) consists of its genus and species, plus any subspecies or variety that applies. The actual name will be italicized, but any clarifying labels will not. For example, the rainbow trout's scientific name is *Oncorhynchus mykiss,* but the coastal subspecies known as the steelhead would specifically be referred to as *Oncorhynchus mykiss* ssp. *irideus.* The "ssp." (sometimes "subsp.") is short for subspecies, and some people don't use it, instead calling this fish *Oncorhynchus mykiss irideus.* However, the abbreviation for variety, "var.," is generally used to help show that the third name is a variety and not a subspecies; *Amanita muscaria* var. *guessowii* is a yellow variety of the classic red and white fly agaric mushroom. There are even cases where multiple varieties of one subspecies exist, in which case both abbreviations are used; *Achillea millefolium* ssp. *millefolium* var. *rubra* is a variant of a subspecies of yarrow that grows in the Appalachian region of the eastern United States.

Sometimes you may also see higher taxonomic names abbreviated with only the more specific name unshortened. *Achillea millefolium* may be seen as *A. millefolium;* the subspecies I mentioned can be written as *A. m.* ssp. *millefolium,* and if you want to save time referring to the variant of the subspecies, just call it *A. m.* ssp. *m.* var. *rubra.* While I'm on the topic of abbreviations, "sp." means a single species, while "spp." refers to multiple species, most commonly in the context of several species in the same taxonomic group. So, if I am talking about a single, undefined species in the genus *Canis,* I might write it as *Canis* sp. But if I want to talk about several species within that genus and not list them all out, I'd use *Canis* spp. (don't confuse it with ssp. for subspecies!).

While taxonomy can be a complicated set of categories full of layers and qualifiers, as with anything in nature identification, it gets easier to understand the more you work with it. If you're feeling overwhelmed, just start with genus and species, since those are the two you'll encounter the most. Practically speaking, knowing the difference between subspecies and variants is the next important step. After that, you can then delve into the rest of the seven common ranks, along with clades, superorders, and other more specialized terms.

Until recently, taxonomy was primarily achieved by examining an organism's morphology—its physical characteristics—and then grouping it with organisms that had a shared morphology. For example, members of the woodpecker family (Picidae) are all birds with straight, tapered beaks; exceptionally long, bristled tongues that they use to forage for insects; zygodactyl feet in which two toes face forward and two face backward; and a relatively slender, compact body that has an upright posture and stiff tail when perched on the side of a tree. While a brown creeper (*Certhia americana*) may look a little like a woodpecker clinging to a tree trunk with its similarly stiff tail, its beak is curved, it lacks the long tongue, and its anisodactyl feet have three forward-facing toes and one backward. These differences in traits are enough to classify the brown creeper in a different family, the Certhiidae. In fact, creepers and woodpeckers don't even share an order; you have to go all the way back to the class Aves to find commonality between these birds.

With the advances in genetic technology in recent decades, scientists have been able to further refine taxonomic relationships. The red-banded polypore (*Fomitopsis pinicola*) was long thought to grow throughout temperate forests in the Northern Hemisphere. However, a 2019 study of the genetics of this fungus found enough evidence to prove that only individuals in Europe and Asia were *F. pinicola* proper. Those in North America were split into three species: *F. mounceae, F. schrenkii,* and *F. ochracea.* Not only are the three species found in different ranges, but each also has a genetic makeup that is sufficiently unique to make it its own species, rather than a subspecies of *F. pinicola.*

We didn't always have access to this wealth of information all the way down to the molecular level. So how did naturalists in the past classify and categorize the many living beings on Earth? Let's look at some of the highlights along the path that got us to where we are today.

Ancient Greece and Rome

Arguably the first Western naturalist whose work has survived to this day is Aristotle. Born in 384 BCE, during Greece's Classical period, he was one of the foremost philosophers of his lifetime. This was an age of great intellectual growth and exploration, and philosophy was a wide-ranging field that focused on asking questions and seeking answers about humanity and the world at large. Philosophers studied and debated everything from natural history to astronomy to mathematics.

Aristotle's works include several natural history texts such as *Historia Animalium* (*The History of Animals*), *De Partibus Animalium* (*The Parts of Animals*), and *De Generatione Animalium* (*The Generation of Animals*). These texts compiled his detailed observations of the living beings he encountered, particularly during his time on the island of Lesbos. He described more than five hundred individual animal species and what traits made each unique, both in body and behavior. He was able to separate characteristics that were less helpful in classifying animals into general groups, such as color and size, while highlighting traits such as the shapes of limbs and other parts, reproduction type, and in the case of animals that metamorphose at various stages of their lives, how those stages play out. And he delved deeper into the origins of animals than did his teacher Plato, who surmised that animals were merely transformed humans whose flaws caused them to devolve.

That said, Aristotle was an imperfect scientist. While his observations did lead to some good inferences about the collective traits of groups of animals, he fell prey to human bias. His work, as well as Plato's, supported the *scala naturae*—the ladder of nature—upon which humans are placed at the top rung, superior to all other beings. And there's a good chance that some of his understanding of vertebrate anatomy came not from previous scientific works, but from hieroscopy, the practice of divining using the entrails of a sacrificed animal.

Still, his work was influential, and among his students was Theophrastus, whom he met on Lesbos. Whereas Aristotle's surviving work concerns animals, Theophrastus's extant writings are distinctly botanical—although given that some of their works have been lost, it is likely that both of these nature-obsessed philosophers wrote about plants and animals alike.

Theophrastus's most notable botanical work is *Historia Plantarum* (*Enquiry into Plants*). This was a colossal ten-volume work, although one that has since been lost, and he may still have been revising it when he died in 287 BCE. Like Aristotle, he based his findings on direct observations, and part of the work involved classifying plants into various groups such as trees, shrubs, and herbs. He even mentioned some of the contemporary theories of plant reproduction:

> *Anaxagoras says that the air contains the seeds of all things, and that these, carried down by the rain, produce the plants; while Diogenes says that this happens when water decomposes and mixes in some sort with earth.*

However, Theophrastus deemed the presence or absence of flowers to be more relevant to a plant's mode of creating seeds and was essentially the first known naturalist to divide plants into what we know today as angiosperms (flowering plants) and gymnosperms (nonflowering, cone-bearing plants).

Pedanius Dioscorides was a physician between the years 50 and 70 who was also interested in plants, specifically for medicinal purposes. He wrote *De Materia Medica* (*On Medical Material*), a text that covered the uses of six hundred plants, as well as some animals and minerals. The book had effective enough medical information that it remained a mainstay of both Western and Arabic medicine for more than a millennium after it was written. While the illustrations could be altered in each copy depending on the transcribing artist, the written summaries of each species were accurate enough to keep the book relevant regardless of what language it was written in. *De Materia Medica* was divided into five sections, each including various plants according to medicinal use and type. This was not taxonomy as we think of it today; instead it was an organizational system based on practical uses of the various species included.

Ancient Rome was known for borrowing heavily from Classical Greek culture and technology, and this includes natural history. *Naturalis Historia* (*Natural History*) is Pliny the Elder's one surviving work, and in addition to animals and plants, he covered topics ranging from astronomy to sculpture. Much of the material he drew on during his lifetime in the first century came from earlier sources, rather than his own observations; instead of presenting new ideas, he essentially created the first encyclopedia.

Pliny the Elder was not the most discerning author when it came to scrutinizing his sources. *Naturalis Historia* famously (or infamously) includes a menagerie of fantastic beings that originated from rumors and may have increased in oddity as the information passed from one person to the next, much like a big game of Telephone. Along with the dog-headed men (Cynocephali) and the monopod people who hopped along on one giant foot (both drawn from the questionable writings of the fifth-century BCE Greek historian Herodotus), there were scaly marine equivalents of several terrestrial animals like sheep and elephants. Mythological beings familiar to us today, like mermaids, unicorns, and the immortal phoenix, made appearances alongside lesser-known fauna like the wildebeest-like catoblepas, who, with a glance, could turn anything to stone, as well as the fiery-boweled bonnacon whose superheated manure and flatulence were expelled at enemies like modern-day napalm.

Moreover, Pliny's organization of the world once again places humanity on top, with the rest of nature arrayed according to its usefulness to our species. The closest he gets to modern taxonomy is, like Aristotle, categorizing nonhuman beings as animal, vegetable, or mineral in nature. It is uncertain how much further he might have developed his ideas, as he died in the year 79 during the eruption of Mount Vesuvius, and most of the *Naturalis Historia* was published posthumously by his nephew, Pliny the Younger.

Pliny the Elder wasn't the only naturalist with an infamous death. The Greek philosopher Pythagoras's death could be seen as a cautionary tale against pseudoscience. One story goes that Pythagoras and his students were being attacked by their political enemies; while fleeing, they came to a field of fava beans. Because they may have believed the beans contained the souls of the dead due to their fleshy appearance and resemblance to human embryos, they refused to enter into the field, and so were captured and slaughtered.

Medieval Bestiaries and Herbaries

The Greek and Roman philosophers, for all their inaccuracies and biases, were attempting to create protoscientific studies of the natural world. Pliny the Elder could be described as something of a pantheist, but that was about as close as this group of writers got to incorporating religion into their works. The same is not true of medieval bestiaries and herbaries, which were compiled and transcribed by Catholic monks, some of whom were accomplished naturalists themselves. In addition to observed natural history, these texts overtly wove Christian theology into their descriptions of various beings both real and fantastic.

An early forerunner of these medieval texts was the *Physiologus* (*The Study of Nature*). Written by an unknown author in Alexandria, it likely dates back to the third or fourth century. It is by no means a complete treatment of the North African fauna of the time, instead covering merely a few dozen animals and a handful of plants and stones. Rather than representing a treatise of pure natural history, it instead consists largely of superstitions and religious allegories with animals as the main characters.

When one looks back at these texts through the lens of modern zoology, some of the claims are obviously fictitious. Lion cubs are said to be born dead, and then on the third day (a hearkening back to the resurrection of Christ), the father blows breath into them. Foxes supposedly play dead to attract scavenging birds, and when they come close enough, the fox attacks them and thereby gains a meal. And there are no greater mortal enemies than deer and snakes; should a deer chase a serpent into its hole, the deer will carry water in its mouth back and forth until it floods the hideout, killing its reptilian foe.

Other entries appear to describe legitimate observed animal behaviors but misinterpret what's actually happening. The *Physiologus* describes how partridges steal other birds' eggs and hatch them, only to have the fledglings return to their own species. This sounds much like nest parasitism, except it is birds of other species who surreptitiously lay their eggs in the nests of the partridge rather than a case of egg theft. Ring-necked pheasants (*Phasianus colchicus*), for example, were introduced to Western Europe during the Roman Empire and are notorious nest parasites of other ground-nesting birds such as partridges.

The *Physiologus* is also where the pelican became a potent symbol of bloodletting. According to this bestiary, at a certain age, her young become quite cantankerous, slapping her and her mate in the face until the parents kill them in a fit of rage. After the requisite three days, she tears open her breast and feeds them her blood to bring them back to life. In reality, young pelicans reach into a parent's mouth for regurgitated fish, and as the older bird's lower jaw may tuck up against its chest during this process, it might look to an unskilled observer—or one at a great distance— that the young are feeding directly on its flesh and blood. Later bestiaries would eliminate the adults' filicide and instead present the female pelican feeding her blood to her young as a sign of self-sacrificing piety.

The *Physiologus* was instrumental in the formation of later bestiaries, which, in addition to African animals, also included European fauna and mythological beings. Dragons, unicorns, and griffins romped through these illuminated manuscripts alongside bears, eagles, and elephants. Medieval illustrations were prone to particularly playful, graceful stylization, although artistic license could give way to anatomical inaccuracies, such as elephants with malformed or nonexistent ears or a hyena with a horse's head and mane. The animals might be presented in alphabetical order or arranged according to whether they lived on land or in water, but otherwise were not organized in any taxonomic fashion.

This image of an owl being mobbed by smaller birds came from the British Library's Harley MS 4751, dating back to the late twelfth or early thirteenth century. The behavior is quite common in real birds, as songbirds and corvids will gang up to chase away owls and other predatory birds. There is not, however, any owl in the history of avian dinosaurs that has had a mouth quite like this one.

The Biblical cataloging of animals and plants in Genesis was taken by most, even among naturalists, to be the supreme order of nature. The closest we might get to an early medieval attempt at the scientific classification of animals is the *Etymologiae* (*Etymologies*) penned by the bishop Isidore of Seville in the seventh century. Like Pliny the Elder's *Naturalis Historia,* the *Etymologiae* was an attempt to catalog as much knowledge of the time as possible, from law and religion to geography and ships, and of course *de animalibus*—the animals. As the text's title suggests, the bishop was especially interested in words, and so he wrote what is essentially an encyclopedia focused specifically on the origins and meanings of vocabulary pertinent to the various subjects he covered.

So it was that Isidore defined individual animal species through the etymology of their names. Some of these were descriptive in a straightforward manner. A type of asp known as the hypnalis (from *hypnos,* "hypnosis"), for example, was known to use its venom to "put to sleep" its prey. Slugs (*limax,* "slime" or "mud") were so called because they were believed to spring forth from mud, or at least were rather dirty, messy animals. And the rhinoceros's name, of course, translates to "horn on the nose."

At times Isidore's conclusions about the origins of an animal's name were correct, but it could be paired with further information that was fanciful at best. Leopards were supposedly born of a female lion (*leo*) and a male panther (*pard*), hence the combined name for the resulting offspring. His entry on deer and their relationship to snakes is a direct callback to the *Physiologus:*

> Deer (cervus) are so called . . . from their horns. They are
> antagonistic to serpents; when they sense themselves
> burdened with infirmity, they draw the serpents from
> their caves with the breath from their nostrils, and
> having overcome the malignancy of the poison, the
> deer are restored to health by eating the serpents.*

---

*   Stephen A. Barney, *The Etymologies of Isidore of Seville,* translated by Stephen A. Barney, W. J. Lewis, J. A. Beach, and Oliver Berghof (Cambridge: Cambridge University Press, 2006), 248.

What bestiaries were to animals, medieval herbals were to plants. As with earlier botanical works like *De Materia Medica,* though, they were largely concerned with the medicinal purposes of plants, and so any scientific organization took a backseat to these practicalities. Some were direct translations of older works; the *Old English Herbarium* manuscripts such as the British Library's Harley MS 585 are tenth- and eleventh-century translations of the *Herbarium Apuleii Platonici (The Herbarium of Pseudo-Apuleiu)* dating back to the fourth century. Others were original medieval manuscripts; *Bald's Leechbook* is a particularly well-known example that, in addition to many herbal remedies, discussed the use of leeches and maggots in medical treatments.

Some of the medical treatments in medieval herbals may stand up to twenty-first-century scientific rigor. As one example, Bald's eyesalve, a mix of wine, leeks, garlic, and cow bile described in the *Leechbook,* contains antibiotic properties sufficient to kill the MRSA bacteria in vitro when left in a brass-lined vessel for nine days.

One particularly notable herbal is *De Vegetabilibus (Of Plants)* by Albertus Magnus. A thirteenth-century Catholic bishop who was posthumously beatified, he was a skilled scientist focused on the process of inquiry. Like earlier naturalist writers, he had diverse interests like logic, phrenology, and alchemy, among many others. Well-versed in his predecessors like Aristotle, he also expounded upon their works with his own ideas and delved into little-studied areas such as mineralogy.

De Vegetabilibus was Magnus's primary work on botany and medicinal plants. The material was based mostly on his observations of his sources, such as the works of Nicolaus of Damascus dating back to the first century, although he drew inspiration widely from Greek, Roman, and Arabic botanical and medicinal libraries. While one section of the text dealt with medicinal properties and another with agriculture, the bulk of it was an attempt at describing the botanical history of various plants. Some of this effort fell short; he wrote, for example, that if you sow ryegrass on sufficiently healthy soil, it would turn into wheat, a "fact" that could be right at home in the more questionable portions of the works of Aristotle or Pliny.

While the strange and often amusing inaccuracies of medieval scientific writing (and earlier source material) often take center stage, it is important to remember that these were legitimate efforts to try to categorize the world and its many beings in a meaningful, deliberate way. The various texts still held valuable content, whether that was solid observations about animals, medicinal properties of plants, or an author's exuberant expression of appreciation for the natural world.

The Next Great Renaissance

As the world seemed to grow larger, so did the thirst for knowledge about it. Existing texts about natural history were joined by new ones uncovered in far-off lands, and once these had been seemingly exhausted, European scientists set about on their own independent studies. Over time, they developed what we now know as the scientific method, a carefully planned process for creating and testing a hypothesis and then examining the results. This was the time of such notable scientists as Galileo Galilei, Nicolaus Copernicus, and Tycho Brahe, along with famed polymath Leonardo da Vinci.

The illuminated manuscripts of the medieval period, with their stylized artwork, began to give way to more realistic illustrations in natural history texts. The Renaissance ushered in greater curiosity about not just the natural world but also how it worked in microcosm and macrocosm alike. And that world seemed to be well-nigh infinite as European ships voyaged across the waves in all directions to exploit the resources of the Americas, Africa, and Asia.

A wide variety of living and preserved organisms brought to Europe through trade gave rise to the cabinet of curiosities, or *wunderkammer* ("cabinet of wonders"). These were no mere freestanding cases, but entire rooms and buildings dedicated to housing and cataloging natural history specimens, cultural artifacts, and art from around the world. While the *wunderkammern* were frequently used as status symbols, with wealthy men vying with one another to have the most exotic and rare collection in the land, they also offered naturalists a greater breadth of life on Earth to explore. In fact, the "curiosity" of the cabinet was often in the mystery each unnamed species offered: something new and unknown to be unraveled, examined, and categorized.

In addition to detailed notes, naturalists often drew or painted detailed replicas of the specimens they studied or hired artists to illustrate their findings. This included depictions of animals and plants in their native habitats, but it also extended to intricate anatomical diagrams. A study of a deceased animal specimen, for example, might include images of the entire corpse, then the flayed musculature, perhaps also some explorations of individual systems like nerves or veins, and finally the skeleton, whether as loose bones or an articulated whole.

Comparative anatomy gave naturalists and other scientists an opportunity to better understand how two or more species might be related to each other, with physical similarities providing the relevant evidence. French naturalist Pierre Belon was a particularly notable figure in the early history of this field, writing several books on subjects such as birds, fish, plants, and more. An illustration in *L'Histoire de la nature des oyseaux* (*The Natural History of Birds*) showing the similarities between the skeletons of a human and a bird is thought to be one of the first diagrams of comparative anatomy.

In comparing species, Volcher Coiter, a sixteenth-century Dutch anatomist, placed particular emphasis on osteology, the study of animal bones. And his body of engravings and other art are exquisite examples of Renaissance-era scientific illustration at its finest. While birds were of particular interest to him, he also made numerous studies of mammals and reptiles. In spite of the continuing social and religious dominance of the *scala naturae* that placed humans at the pinnacle of creation, he went so far as to boldly illustrate how the skeletons of humans closely resemble those of our primate cousins.

By comparing and contrasting the shapes of the bones themselves, as well as the overall skeletal structure, Coiter proposed a series of classifications for various groups of vertebrate animals. He even created a forerunner of the modern dichotomous key for birds, in which anatomical features would be presented in a pair of alternatives, with each selection leading to another pair until one finally arrived at a species that fit all of the chosen features. At a time when naturalists were working to create a scientific order of the world, Coiter left an indelible mark by pioneering comparative osteology and offering a clear way to categorize animals based on anatomical structures.

Other contemporary naturalist-artists like Hieronymus Fabricius, Nicolaus Steno, and Samuel Collins continued to produce written and illustrated studies of various elements of animal anatomy, with the growing inclusion of the human animal along with the rest. One of the greatest European Renaissance attempts at a universal cataloging of life-forms came late in the period, care of John Jonston along with the keen artistic talent of Matthäus Merian (the elder). Together they produced the *Historiae Naturalis* (*The Natural History*), a series of books describing and organizing groups of animals like snakes, mammals, and fish. Even insects and other invertebrates were examined in these labor-intensive works, and animals of both the land and sea were described in great detail with finely crafted illustrations.

While Jonston's writings were of good scientific caliber for the time, like other naturalists, he sometimes had to rely on secondhand knowledge. To wit, *Historiae Naturalis de Quadrupedibus* (*The Natural History of Four-Footed Animals*) features illustrations of several unique species of unicorn. A background image showing a unicorn lured into a trap baited with a fair maiden is quite the callback to medieval bestiaries!

The Renaissance was a time of great tension. Whereas in medieval times, monks and other clergymen were among those pursuing the natural sciences, more recent practitioners often found themselves at violent odds with the church. Galileo spent his last decade under house arrest after defending Copernicus's claim that the Earth orbits the Sun, and both Giordano Bruno and Michael Servetus were burned at the stake.

In this crucible was born the Scientific Revolution, which saw not only great advances in the sciences but also in the very way European people viewed the world. This movement would bridge the Renaissance with the Enlightenment. The sociocultural, political, and religious revolutions that had first taken seed in the medieval period and grew throughout the Renaissance would now take full flight in the Age of Reason.

The Enlightenment and Beyond

Had Copernicus and Galileo been able to stick around a few more centuries, they would have been gratified to see heliocentrism become the accepted model of the cosmos. But this development was par for the course at the time. By the eighteenth century, naturalists and other scientists had been collecting, sorting, and examining species of organisms from around the world in greater detail than ever, and it became increasingly evident that the world was much more complex than anyone had imagined. An ever-larger number of fossils challenged the idea that the Earth was only six thousand years old and that the only species that had been created were those currently in existence. Moreover, the increased understanding of comparative anatomy, including that of *Homo sapiens,* clearly showed that we could easily be counted among the other animals that walked upon this Earth. Empirical evidence and the scientific method were in; Aristotle, Genesis, and the *scala naturae* were out.

This meant that new methods for organizing living beings were needed, and the groundwork laid by Renaissance naturalists like Coiter and Jonston created the necessary foundation for their successors. John Ray provided one of the earliest Enlightenment-era attempts at categorization. As a parson-naturalist, he felt his explorations of natural history were necessary to truly understand God's creation, and he proceeded to not just understand but organize it as well.

One of Ray's best-known works, *The Wisdom of God Manifested in the Works of the Creation,* combines direct observations of animals and other beings with religious allegories. In some respects, it very much resembles the medieval bestiaries in its blending of science and religion; for example, when speaking of the bodily proportions of various animals, he chalks it up to evidence of God's design that they are all perfectly formed for their tasks in finding food, moving about the land, and so forth.

But perhaps Ray's most important contribution to modern science was his method of classifying organisms. Rather than using dichotomies as Coiter had, Ray instead determined that animals, plants, and other beings could be differentiated from one another as distinct *species.* He went on to organize these species into larger groups based on physical characteristics. His botanical works went into especial detail and provided a system that resembles modern taxonomy more than any previous catalog.

Ray started by classifying all herbaceous plants as Herbae, and then into the subcategories Perfectae (seed-producing plants) and Imperfectae (spore-producing plants). The Perfectae were then further divided into monocots, whose seedlings first sprout with only one leaf, and dicots, which have two-leaved seedlings. Trees (Arborae) were separate from the Herbae but were also divided into monocots and dicots. These were the basic categories used in his *Historia Plantarum* (*The History of Plants*), but an earlier work, the *Catalogus plantarum circa Cantabrigiam nascentium* (*A Catalog of Plants Growing around Cambridge*), went into even more detail by dividing the Herbae into almost two dozen other groups. We can see the resemblance to modern plant taxonomy in categories like the Bulbosae (plants that grow from bulbs) or the Succulentae (*Sedum* and other succulents).

After Ray's death in 1705, naturalists and other scientists discovered and debated evidence that showed the Earth's age was much older than the six thousand years purported by Christianity. This growing realization of the depth of time and the mutability of life challenged the understanding of planetary history both within and outside of the scientific community.

It is from this point that we begin seeing the motif of the scientific tree of life. Augustin Augier, a French botanist and Catholic priest, is thought to have been the first to arrange organisms according to relationships on a tree-shaped diagram. Jean-Baptiste Lamarck, almost a decade later, produced an animal tree of life in his *Philosophie zoologique* (*Zoological Philosophy*). Lamarck famously posited that the charted relationships pointed to an evolutionary history; although rather than multiple species originating from the same ancestor, he instead thought that multiple independent animal lineages all proceeded from simpler to more complex forms.

Perhaps the best-known figure in the field, the Swedish botanist and zoologist Carl Linnaeus took categorization to new levels of complexity. One can see the influences of past naturalists in his immense undertaking; for example, his *Systema Naturae*

(*System of Nature*) brought forth the classification structure we are familiar with today. His three initial kingdoms—Animalia, Plantae, and Lapideum—call to mind the "animal, vegetable, or mineral" of the ancient Greek philosophers. He further divided each kingdom into classes, orders, genera, and species. And by the 1758 edition, he had begun using binomial nomenclature, referring to each being by its genus and species.

Linnaeus's work was so fundamental that even today we still refer to "Linnaean taxonomy." However, the system has been much expanded and revamped since his initial formulations. We no longer include minerals, for example, and the two remaining kingdoms have since been joined by anywhere from three to six others, depending on who you talk to. The domain has become the highest taxonomic rank, and Linnaeus's initial four subdivisions are further diversified with phyla and families, tribes and clades, and the assortment of subspecific ranks.

The theory of evolution that was hinted at in Enlightenment-era explorations exploded with the publication of Charles Darwin's *On the Origin of Species* in 1859; since then, evolutionary biology continues to play a major role in updating and refining taxonomy. Advances in genetic research allow scientists to determine how individual species are related—or not—based on their DNA.

Taxonomy is not set in stone. Species frequently do not have the same scientific name they had when they were first described by science, and sometimes entire taxonomic groups are renamed. The continual discovery of species new to science requires the reshuffling of various family trees, even when the species is nestled within a well-established taxon rather than being a monophyletic species all alone in its genus or family. Well-known species may also end up getting moved to new taxonomic locations as we learn more about them, and they are sometimes reclassified multiple times in the span of a few years. And species that don't seem to have a neat niche within a current classification scheme often end up relegated to a wastebasket taxon along with other assorted misfits with whom they may have only tenuous relationships.

Common names can also get complicated. *Puma concolor,* for example, is known by more than forty English names, including cougar, puma, and mountain lion, and there are even more in various indigenous languages. Sometimes species are deliberately renamed, as in the American Ornithological Society's 2023 decision to give all birds named after people more descriptive monikers based on characteristics like color. There are also cases where a common name is misapplied; the blue jay (*Cyanocitta cristata*) is rarely found west of the Rockies, but people on the West Coast sometimes erroneously refer to Steller's jays (*Cyanocitta stelleri*) and California scrub jays (*Aphelocoma californica*) as blue jays.

If this all seems confusing, don't feel bad. Even the people responsible for classifying organisms often have trouble keeping up on the who's who of species in this constantly changing field. Changes in taxa are sometimes met with passionate debate, proof that scientists can, in fact, be quite a lively bunch.

But this process of research, definition, and updating is a tradition that goes back thousands of years in Western science. As you learn the common and scientific names of species, over time you'll find that some of these names may change in your field guides and other reference materials. The best advice I can give you is to simply go with the flow, learn the new names, and, like any good naturalist, always be open to new information about the beings with whom we share this world.

CHAPTER 3

Getting Started with Identification

Oregon dark-eyed Junco (*Junco hyemalis oreganus*)

I WROTE THIS BOOK FOR EVERYONE I'VE EVER MET who, despite access to field guides, dichotomous keys, ID apps, and websites, had no idea where to start in using any of them to identify something they had found. It's one thing to have access to a resource, but another to be able to competently apply it. There's also a wide gap between "I have absolutely no idea what I'm doing with any of this" and "I am a professionally trained field biologist." Chances are you're somewhere in between those two extremes. So, before we embark on how to identify individual kingdoms of organisms, I want to introduce you to some of the tools you might find handy in nature identification, practical considerations before you head out into the field to find cool stuff, and a breakdown of my own process in figuring out what animal, plant, fungus, or other organism I've just met for the first time.

I do want to emphasize that every person's approach to nature identification is going to be different. Some people are good at observation and paying attention to important details of a given organism's appearance, behavior, and so forth, while others may have a tougher time discerning what makes this plant or mushroom different from any others they've encountered before. There are those who absolutely love dichotomous keys and many who simply can't make the keys work for them. While I'm going to be outlining the way I do things, I encourage you to figure out what tools and methods work best for you.

There are a couple of caveats, of course. First, never rely on only one tool for identification, especially if that tool is an ID app. Any of them can be used incorrectly or have misinformation, and so cross-referencing with multiple resources is an absolute must. I've heard some people say, "If you get three books to agree on something, then it's true." I'm a little more cautious than that, which means I am going to check every single book I have, plus run any photos I have through at least one app and check multiple websites. Depending on how confident I am (or not) in my identification, I often consult other everyday naturalists, in person or online, to see if they agree.

Second, if you are consistently having trouble getting solid identifications, there's a good chance you need to do something different. Maybe the field guides you have access to don't have the right information or aren't organized in a way that makes sense to you. Or perhaps you need to switch up what order you bring in different resources. It's okay to try a variety of possibilities to see what helps you be more confident in your ability to correctly identify an organism—just make sure you give each new resource you try enough time for you to become familiar with it and how it works before you give up on it entirely.

Once again, I remind you that nature identification is a skill that takes practice, so you may just need to keep at it until you get better. There are at least as many ways to get it wrong as there are to do it right, and you're bound to make mistakes along the way. But that's a part of learning anything new, and if you don't do it perfectly the first time, it doesn't mean you're going to be terrible at it forever. I wouldn't be writing an entire book on identification if it really were as simple as "Open the app, take a picture, and voilà—there's your answer!" So be patient with yourself, and look at this learning experience as a marathon, not a sprint.

Field Guides

A field guide is a book (or occasionally a pamphlet) that contains names, pictures, range maps, and other identifying characteristics of various species. Some field guides are general, like an overview of all the bird species in North America. Others explore the animals, plants, and fungi of a particular country, state, or other region. Some field guides are more specific in their scope, whether they focus on a narrow range of living beings, a relatively localized area, or a combination of both. You can even find field guides on bird feathers and animal tracks.

The information in each species' entry may vary depending on the scope of the field guide; one on birds, for example, will often include some typical behaviors of each species, while a guidebook on fungi may mention whether a given mushroom is edible or toxic. Not every field guide is going to be as thorough as others; a pocket-sized guide will necessarily be briefer so that it retains easily portable dimensions. If you really value having a more comprehensive text with you, be prepared to carry the extra weight in your backpack (or plan on buying and downloading some ebooks).

Because I collect field guides, I have a terrifying number of them taking over my shelves. You likely don't need that many (neither do I, probably), but it is good to have access to at least a few different guides when possible for each type of organism you'd like to identify. Birders and mushroom hunters have the luxury of a wide variety of books both general and more specific, while the only field guide to slime molds I have in my collection is *Myxomycetes: A Handbook of Slime Molds* by Steven L. Stephenson and Henry Stempen. Many field guides focus only on native species, which means if you're in an area that has many invasive or other nonnative species, they won't necessarily be in your books. Some regions of the world have more field guides to their flora, fauna, and fungi available than others; I'm spoiled for choice here in North America, but someone in rural Mongolia will have far fewer options.

Let's look at a sample entry that might be found in a field guide to everyday naturalists, if such a thing were ever written:

NAME: Rebecca Lexa (*Homo sapiens* var. *naturalista*)

HEIGHT: 5 feet 3 inches

DESCRIPTION: Pale skin, although may have darker "T-shirt exposure" markings on the face, neck, and arms, especially in summer and early fall. Crest of light brown hair usually folded at the nape, but may be raised with alarm, excitement, bed head, or absence of a hair scrunchie.

RANGE: Pacific Northwest US, particularly the Columbia-Pacific region, although may be found as an occasional vagrant as far east as Missouri.

DIET AND BEHAVIOR: Omnivorous, although diet consists primarily of plant matter and cheese. May be enticed into the open for observation with chocolate and other sweets. Most likely encountered on trails throughout the Pacific Northwest, whether in isolation or as part of a group; tends to be loquacious in the presence of other *Homo sapiens*. Has various dens scattered throughout the region and, in such settings, is often found in the vicinity of a computer.

SIMILAR SPECIES: For the good of all of us, let's hope not; one is more than enough!

So how do you choose which field guides to have on hand? Start by heading to a library or bookstore and browsing the field guides there. That will help give you a feel for what's available and what makes each book unique. People often end up having preferences; I tend to prefer more specific field guides that explore regions or groups of beings because they tend to include most, if not all, of the expected species in their purview. A general field guide can be useful for getting started or taking out into the field for casual identification when you have to keep your backpack light, but there's a very good chance you're going to come across something that isn't included in its pages. I do keep some general guides on hand for comparative purposes when I'm researching a particular species, as they still have good information on the species they do cover.

Here are some things to consider when looking at field guides:

Is It General or Specific?

As mentioned above, a general guide to the flora, fauna, and fungi of a large area won't have every organism found there, but it will include many of the common species. You also won't have to swap back and forth between a book on shorebirds and then a guide to gilled mushrooms and then one about salamanders in the southeastern US. If your book budget is limited or you really want only one book to start with, go with a general guide.

That said, there is value in more focused texts. That interesting newt may not be in your general field guide, but it gets a thorough entry in the Southeastern salamanders book. Additionally, if you especially like looking for and identifying certain groups of organisms, field guides that concentrate on them usually have more detailed information. Don't assume that two specialized field guides on similar topics will be exactly the same, though; some field guides haven't been updated in a while and may be out of date, while various authors may disagree on certain details of a given species.

Does It Have Photos or Illustrations?

Some people really like photos because they're true to life. A sharp, well-lit photograph showcases the species it depicts, clearly demonstrating traits like colors, shapes, typical postures of animals, and so forth. Older field guides may still have black-and-white photos, but these days full color is standard.

Illustrations have advantages, too. The artist can choose exactly how they want to portray their subjects in a way that best shows identifying characteristics. One of the reasons I love David Sibley's field guides to birds is that he paints each species in the same series of poses—perching, flying, swimming, and so on—with notes that highlight unique colors, markings, and other traits. This makes comparing similar species a lot easier; you can clearly see, for example, that the Bewick's wren (*Thryomanes bewickii*) lacks the reddish tones and yellowish breast and belly of the similar Carolina wren (*Thryothorus ludovicianus*).

If there's enough room, a good field guide will also include photos or illustrations of regional variations of a species, life stages, and so forth. Remember that a smaller pocket guide will have fewer images than a more in-depth book, so if a variety of pictures is important to you, skip the pocket guides.

How Is It Organized?

Field guides on the same topic may be organized differently. Those that cover all the birds of a given area will usually arrange them in the table of contents in commonly used classifications, such as "tyrant flycatchers" or "wagtails and pipits." But one book may have all waterfowl (ducks, geese, and swans) together in one category, while another's table of contents breaks them out into dabbling ducks, diving ducks, and so forth. Most field guide indices include the ability to search for both the common and scientific names of the species included, but some have one common index and one scientific.

Do you have to memorize the scientific name of every species you encounter? Not really. Most of the sources you'll be using also include any common names a species has. The ones that I can easily remember are those I've referred to many times, but I haven't gone through the trouble to deliberately memorize the scientific names of all the species I know. You'll likely have at least a few stick in your mind as you keep practicing nature identification, but don't worry if they don't automatically spring to mind; you can always look them up as needed.

This sort of categorization can seem a bit frustrating if you don't know whether you're looking at a vireo or a warbler, which means you may have to take more time paging through the book to find a bird that looks like the one you've found. As you become more familiar with your local birds—or plants or mushrooms—the taxonomic categories they're divided into start to make more sense. With practice, you'll be able to quickly turn to the most likely section of the book in the search for your mystery species.

Occasionally you can find a field guide that uses a more beginner-friendly organizational system. *Birds of Oregon Field Guide* by Stan Tekiela groups its species by color, one of the most common traits people use to initially describe an unknown bird. Similarly, one of my favorite plant field guides, Damian Fagan's *Pacific Northwest Wildflowers,* organizes plants by the color of their flowers. Within those color sections, species in the same genus or family may be found next to some that are completely unrelated, but starting with a trait that is easy to identify makes these field guides more intuitive for some people. Organization by color works best in books that are more region-specific; once you start getting into books that cover larger areas, such as North

America, the sheer number of species becomes too unwieldy, and more detailed organizational systems are needed. Color also doesn't work so well with fungi, since the color palette is somewhat more limited (there are a *lot* of brown mushrooms).

How Thorough Is the Information?

General guides usually have briefer entries, while more specific ones often go into more detail. Think about what information is most important to you when trying to identify a new species. Do you want the entry to include as many details as possible, like animal behavior or trees that a fungus has mycorrhizal relationships with? Or are you happy with the basics like color, size, range, and a few good pictures, and you'll look up the rest online and in other sources?

How Reliable Is the Information?

Just because a book is in print doesn't mean it's going to have valid information. If you've been studying a particular group of organisms for a while, you may have a pretty good idea of the signs an author has no idea what they're talking about. But if you're new to all of this or you're delving into areas of biology you don't normally explore, you may not have a good frame of reference to gauge the quality of a given text.

This is where book reviews come in handy. Yes, it's easy to just head over to an online bookseller and look at the reviews people have written about a particular field guide to see what they like and dislike. But if you can find more professional reviews like those in newspapers, magazines, and outdoor websites, you may get more detail on the book's pros and cons. Look at the credentials of the review's author; if they have experience in a relevant area of natural history, that means their review probably carries some weight.

Don't dismiss a book just because it has a few bad or lukewarm reviews; no book is going to be universally loved (not even this one), and so it's normal for some people to not be as thrilled by it as others. Just as you would look up a species in many different resources, it's a good idea to compare several reviews of a field guide when trying to determine how good it is.

You'll also want to look up the book's author; if you find a bunch of articles about them making things up, for example, you might want to steer clear of their work. Check out their website and other online presence to see if they list any credentials or relevant experience. Many authors also write articles and other short-form writing; this is a good way to get a feel for their work.

If you're part of an online group devoted to the types of living beings you want to identify, you can ask people there for their suggested reading. You'll likely get a decent number of responses, so make a list of all the books they recommend. Do be aware that when there are longtime favorite titles on a given subject, people may automatically recommend them because that's what *everyone* recommends. Among Pacific Northwest mushroom hunters, David Arora's *All That the Rain Promises and More . . . A Hip Pocket Guide to Western Mushrooms* is the most popular field guide to the edible fungi of the region. It is a solid book from a highly experienced author that deserves its glowing reputation, but it also hasn't been updated since it first came out in 1991. I tend to recommend it along with more current books like *Fruits of the Forest: A Field Guide to Pacific Northwest Edible Mushrooms* by Daniel Winkler because the more current books go into detail on newer information, such as how angel wing mushrooms (*Pleurocybella porrigens*) may not be as edible as we thought, because of the species causing fatal poisonings in 2004 and 2009.

Finally—and I cannot believe how utterly dystopian it is for me to even write this sentence—please beware of AI-generated field guides. With the rise in programs like ChatGPT that will generate bodies of text based on topic prompts, an entire book can be assembled in a matter of minutes. The people who are behind these books have zero experience in the subject matter and don't do any editing for content or writing quality, and so every single one of these texts should be suspect.

This practice is especially common with field guides of edible plants and mushrooms, as foraging is a hot topic with a growing audience. These books are particularly dangerous because no one is fact-checking their claims, even though they are presented as authoritative texts on what is safe to eat and what isn't. In reviewing some of these books, I've found they often omit crucial identification information, misidentify species and their edibility, and, in some cases, even use AI-generated illustrations that in no way resemble the actual plant or fungus.

I've written a guide on how to identify AI-generated foraging field guides in a blog post on my website, RebeccaLexa.com, and the information there generally carries over to other AI books as well. Signs that a book may be AI-generated include obvious spelling or grammatical errors (especially in the title and book description), an "author" with no profile information on the internet, or an "author" who has supposedly published several books on disparate topics within a matter of a few days or weeks.

How Heavy Is It?

This may seem like a silly question, but there is such a thing as taking too many books with you on the trail. For someone who prefers to travel light, a few ounces can make a big difference. And anyone may feel the strain when a pack that felt fine a half mile in starts to feel like a bag of rocks at the five-mile mark.

This is where pocket guides can come in handy, especially if you prefer to have multiple books with you. Some online retailers list exactly how heavy a given book is. If you are choosing your field guides based on their physical weight, it's a great quick reference.

Some field guides also come in ebook format. If you like using ebooks and you feel confident in your phone's battery life, this may be a great lightweight option for you. Just buy the ebook, download it to your phone, and you don't have to haul the extra weight of a paperback with you.

How and When Will You Use It?

You may find that the field guides you take with you when exploring aren't the same ones you prefer to use from the comfort of your home. Maybe you can get away with a pocket guide out on a trail, or a general flora, fauna, and fungi book when camping for the weekend, but then when you get back home, you want to consult more sources on the species you observed. Many naturalists curate a collection of varied field guides.

Public libraries also often have a terrific selection of field guides available for checkout, and while you may be limited in how long you can keep them, you at least have the opportunity to look them over and see which ones you like. It's a great opportunity to try it before you buy it from your local bookstore! If the library doesn't have the field guides you're interested in reading, they may be able to find them through interlibrary loan. And librarians love it when patrons suggest books they'd like to see in the collection. There may be a waiting list for more popular titles; around where I live, mushroom guides are scarce on the shelves during fall mushroom-hunting season, for example.

Dichotomous Keys

True to their either-or nature, dichotomous keys seem to be one of those tools that people either love or hate. A key will lead you down various pathways based on your answers to questions like "Does the specimen look like this or that?" and "Does the specimen have this feature or not?" By the end, you will hopefully come to a species recommendation that matches all the traits you picked. They're sort of like a *Choose Your Own Adventure* book for nature identification, except you choose from the options you're given at each step based on the specimen you're trying to identify.

For example, in B. J. Verts's *Keys to the Mammals of Oregon,* the first key is "A Key to the Orders of Mammals in Oregon." Your first choice is between these two options:

> *1a. Cranium small (Fig. 11); teeth 50, upper incisors five on each side; epipubic bones present; hallux opposable and equipped with a nail; penis bifurcate, female with a marsupium enclosing nipples.*

Or

> *1b. Cranium relatively large (Fig. 11); teeth 44 or fewer; upper incisors three or fewer on each side; epipubic bones absent; hallux, if present, equipped with a claw, and not opposable; penis simple; marsupium enclosing nipples absent.*

Figure 11, by the way, consists of line drawings of the skulls of the Virginia opossum (*Didelphis virginiana,* the only marsupial native to North America) and the raccoon (*Procyon lotor*) as representatives of placental mammals found in Oregon. If you have some skull specimens that you're trying to identify, it's pretty easy to count the teeth to see if there are fifty or forty-four or fewer. However, you're probably not going to be able to get close enough to a live Oregonian mammal to determine whether the hallux (the innermost toe on a hind limb, like our big toe) has a nail or a claw; you might not even know what the difference between a nail and a claw is for the purposes of this key. And forget about investigating whether this animal has exposed nipples or has hidden them in a marsupium (pouch), or, if it has a penis, whether it's bifurcated (forked) or not.

Fun fact time: Opossums' native range in North America consists of the United States east of the Great Plains and most of Mexico. They expanded in the past century because of environmental changes wrought by settlers, and they were brought to the Pacific Northwest early in the twentieth century, likely as pets and as food.

Let's say you were trying to key out lichens. *Macrolichens of the Pacific Northwest* by Bruce McCune and Linda Geiser and McCune's two-volume *Microlichens of the Pacific Northwest* are the only guides dedicated solely to the lichens of this region. *Macrolichens* has a key starting on page 1. These are your first two choices:

> *1a Fungus a basidiomycete; fruiting structures mushroom-like or club-shaped.*
>
> *1b Fungus an ascomycete; fruiting structures various but not mushroom-like.*

Those fruiting structures are microscopic, so you won't be able to determine what they look like with a casual glance. If you happen to know that a given lichen's fungal partner is a basidiomycete (which we'll discuss more in Chapter 7), you would then go to Key A on page 2 of *Macrolichens*, but if it is an ascomycete, you continue down to these two choices:

> *2a Thallus gelatinous, without obvious internal layers, black to brown or gray; photobiont blue-green except in two intertidal spp.*
>
> *2b Thallus stratified (or too finely filamentous to tell), color various; photobiont blue-green and/or green.*

Now we're getting into more technical terminology, like thallus and photobiont; if you're familiar with lichens or fungi in general, you may already know these words, but many people aren't. Moreover, you need to know how to determine whether the thallus has layers, is stratified or not, and how to determine the color of the photobiont (its photosynthesizing partner, generally an alga or cyanobacterium).

Lichens are notoriously difficult to identify down to the species level, even by experts, and unlike plants, we aren't familiarized with the basic parts of a lichen in elementary school. Keys almost invariably rely on increasingly technical terminology and observations as you continue through your choices. You can look up the definitions of unknown words, but often trying to discern the quality of a given characteristic—like a layered or nonlayered thallus, for example—isn't something the everyday naturalist is going to know how to do.

My point is that dichotomous keys can be tough to use, especially if you don't have someone showing you how to use them and what each of those words means. Occasionally an author will try to make a more layperson-friendly key. *A Key to Missouri Trees in Winter* by Jerry Cliburn and Ginny Wallace uses plain, clear language and great line illustrations to help an everyday naturalist key out trees even when all you have access to are leaf and flower buds waiting for spring. Michael Beug's *Mushrooms of Cascadia* contains pictures showing exactly what structures you should be searching for and what they should look like, and it also annotates each step with helpful information. It still takes practice to get used to it, and I wouldn't recommend it as your only tool if you're just getting started with fungus identification, but Beug has put a lot of effort into making it understandable.

A few books compile and define relevant words. *Plant Identification Terminology* by James G. Harris and Melinda Woolf Harris defines pretty much any plant-related word you're going to run into in field guides and other sources, and it features line drawings that show what type of structure each one refers to.

It also has sections dedicated to terms concerning individual parts of a plant, such as names for different features of leaves, parts of flowers, and so forth.

Robert M. Hallock's *A Mushroom Word Guide* does the same thing for fungi, except without the pictures, although the pronunciations are very helpful for those of us who have seen but never heard these words. A good companion, if you can find it, is *How to Identify Mushrooms to Genus I* by David L. Largent and Daniel E. Stuntz. It contains not only a regular key, but also a more expanded one with illustrations of what certain terms are referring to. It's similar to Beug's work, but whereas Beug might show you a photo of a mushroom with a decurrent gill attachment, Sharon Hadley's illustrations zoom in on a cross-section of several different types of gill attachments shown together for comparison purposes, which I feel makes it a better book for getting visual representations of fungus-related terms.

All that said, I personally don't really use dichotomous keys. I'm too impatient, and there are other ways to narrow my options when I try to identify an organism down to the species level. If you want to explore them, though, you may find that they're right up your alley.

Online Sources

The internet has provided an incredible amount of information that is constantly being added to, and it has made learning about nature more accessible than ever. Literally anyone with an online connection can create a website, join social media, and add to the ever-growing bank of knowledge out there. But just because it exists on the internet doesn't mean it's true. Being able to assess the veracity of a given online source is yet another crucial skill needed for solid nature identification.

First, consider the website in question. Do the people running it have experience and credentials of any sort? Take AllAboutBirds.org, one of my favorite websites for information on identifying North American birds. It's run by the Cornell Lab of Ornithology, which is about as reliable as it gets when it comes to our avian neighbors. Generally speaking, websites that are managed by universities and other academic organizations are going to be pretty solid, as are well-regarded newspapers, magazines, and other journalistic sources. Look for the sites of nonprofit organizations that focus on various groups of animals, plants, or fungi, such as the North American Mycological Association or your local mycological society, Bird Alliance chapters, herpetological organizations, and native plant societies, as their boards and many members will be experts in their respective fields. National Wildlife Refuges, national and state parks, and other governmental natural resource departments often have good information about local flora, fauna, and fungi on their sites, too.

Just because a website is managed by an individual rather than an organization doesn't necessarily mean it's going to be of lesser quality, but do check its author bio or about page to learn more about who, exactly, is providing the information on the site. MushroomExpert.com is the creation of Michael Kuo, a self-described amateur mycologist (with some significant mycology-related publications) who has maintained the site since 2000. Mark Turner is the photographer behind the thousands of photos that make PNWflowers.com such an amazing visual reference; he has also cowritten multiple field guides to northwest flora. HerpsofArkansas.com is a guide to identifying the reptiles and amphibians ("herps" for short) of Arkansas. It's run by Kory Roberts, who also produced the *Arkansas Herpetological Atlas 2019* with the support of the Arkansas Game and Fish Commission.

A person can have a lot of knowledge about their particular realm of the natural world without being a published author. Another good way to help gauge the quality of a website is to look at how reliable the information they provide is. Do they cite references, such as a bibliography at the end of an article or websites linked throughout the text? You can use the material in this section to assess the quality of those online sources. Pick a few articles or posts on the site and fact-check them; if what they say seems to be in line with other articles on the same topic, that's a good sign of reliability.

In some cases, the information on a website will have been crowdsourced from a variety of people rather than written by one or a few authors. HerpMapper.org allows people to upload pictures of amphibians and reptiles from around the world with other identifying information like location, age, and whether the animal was alive or deceased. While there is not someone specifically checking the veracity of every single claim, this site's users tend to be herp enthusiasts with a lot of accumulated knowledge. If you're searching the site to find pictures of a particular species, a misidentified photo will often stand out from the rest.

Speaking of reliability, consider any AI-generated online content to be immediately suspect. AI doesn't know how to check the validity of its output; it's merely scraping the internet and then cobbling together an answer based on what it finds, regardless of whether it's true or not.

As with AI-written books, immediate signs that something may not have been written by a human include spelling and grammatical errors, strange choices of words that don't flow well (for example, "Based on biology evidence" instead of "Based on biological evidence"), and repetitive text. However, incorrect or unusual language can also mean the writer is an actual human whose first language isn't English, or they just aren't a polished writer, in which case you'll want to assess the content of their work as you would any other website.

Online groups are the ultimate crowdsourced information. Some of these are specifically geared toward a certain area of nature identification (like reddit.com/r/WhatsThisBug/), while others allow more general on-topic conversation in addition to ID help requests (like facebook.com/groups/TheEntomologyGroup). Whether you're looking at Facebook groups, Reddit subreddits, or other communities, you'll want to exercise some caution. While many communities are moderated, that doesn't mean every incorrect comment will be removed, so don't assume that the first comment you see is correct.

I find scrolling through other people's posts to be a valuable learning opportunity. Not only can I see how they might debate the identity of a given species—or come to a consensus—but over time I can also see who the more reliable responders are based on how often they leave accurate comments. Then if I decide to post asking for verification of the species I observed, I have a better idea of whose answers are likely to be accurate.

I recommend waiting to post until you've done some work on your own. These groups already get many people posting, "Hey, what's this plant, bug, and so on?" You'll give yourself good identification practice without taking the shortcut of asking a bunch of other people first, and group members often appreciate those who are willing to put in some effort. Make sure you follow the group's rules; some may require that you include only one species per post or that you always mention your location.

One other trick I like to use involves search engines. Let's say I'm in Oregon and I see a blue and black bird I don't recognize. I would then open a search engine—I like Ecosia, since they plant trees for every certain number of searches made—and type in a search string with the type of organism I observed, where I observed it, and a distinctive characteristic or two. In this case, I might type "blue and black bird in Oregon." Once I get my search results, I go to the image results and then scroll through to see if a bird appears that looks like the one I saw. If it does, I then click on the website to see if the photo has a caption or other indication of what bird is pictured.

If you're not getting the results you want, you might need to make your search more specific, like "blue and black bird with crest in Oregon" or "blue and black bird at Mount Hood." Conversely, you also may need to generalize it a bit; if "blue and black bird at Trillium Lake" isn't helping, try "blue and black bird Pacific Northwest" (or my original string, "blue and black bird in Oregon").

Check the website where a given picture is from and compare images of the same species from multiple sites; unfortunately, both computer-altered and AI-generated "photos" are increasingly polluting image results. The cross fox is a melanistic version of the red fox (*Vulpes vulpes*); it is gray to black with some patches of yellowish orange similar to the color of a nonmelanistic fox. An image search of cross foxes brings up numerous photos in which the colors have been enhanced with a bolder, almost neon red-orange and a darker, more uniform black than what you would see in real life.

AI-generated images often have a slick, "too real" appearance, like the animal is made of plastic. A closer inspection often shows poorly rendered toes, beaks, horns, or other details; they may be misshapen, blend into other parts of the photo, or there are too many or too few.

A last caveat regarding online sources: They can disappear without warning. You may head to your favorite website or online group only to find it's been permanently taken down, even if you were just there yesterday. If a site you've used has gone offline, it's possible that the site has been archived at the Internet Archive's Wayback Machine (https://archive.org/). If you search the URL of your dearly departed reference site, you may get to see older archived versions of it.

ID Apps

I'm discussing apps separately from the rest of the internet because of how they function. Using your phone, you start by taking a picture (or, in some cases of animals, a sound recording) of whatever living being you're observing (specimen data). You then upload it to the app, which may also record the date and time (seasonal data) and GPS coordinates (location data). Using these three types of data—specimen, seasonal, and location—the app's algorithms then search through a database to find similar-looking or -sounding organisms found at the same time of year in the same area, and then the app generates a suggested species or a list of a few possibilities.

I say this several times throughout the book, but I want it engraved into your brain in great big letters: Never, ever, ever use an app as your only tool for nature identification. Apps are great for getting some initial suggestions of what you might have seen, but they are prone to error. If your photo or sound recording wasn't clear enough, if the species you observed has many look-alikes or soundalikes, if the algorithms aren't well designed, or if the database the app draws from isn't extensive enough, you may end up with incorrect suggestions.

I know some of you might be foragers, so my warning about using nature apps as a standalone tool goes triply so for you folks. A hapless naturalist getting a rash from poison oak because an app told them it was white oak is bad enough; once you start ingesting your finds, you ABSOLUTELY must be sure they're edible species. Please don't add to my collection of news stories about people who ate poisonous plants or mushrooms after an app told them they were edible species.

Moreover, part of being a thorough naturalist involves checking your work. I really love the Merlin app for its live identification of bird songs and calls, but I've had it confuse similar-sounding species before, such as the northern flicker (*Colaptes auratus*) and pileated woodpecker (*Dryocopus pileatus*). If Google Lens tells you that the flower you found is Queen Anne's lace (*Daucus carota*), then you need to confirm that it's not a closely related, highly toxic member of the Apiaceae, such as poison hemlock (*Conium maculatum*) or some species of water dropwort (*Oenanthe* spp.). If an app gives you several potential species options, then you'll have to research each one and determine which is the best match for the being you've observed. It won't always be the first one on the list, and sometimes none of them are a clear match.

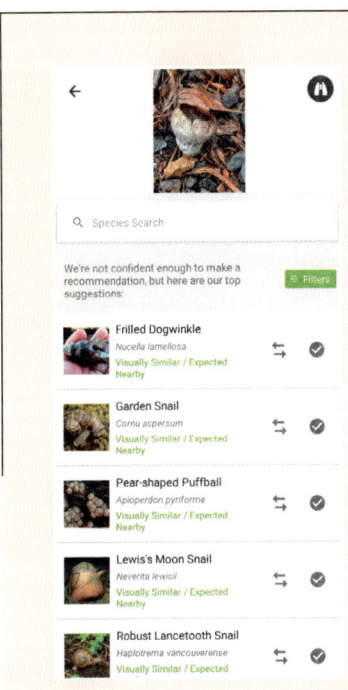

I think it's fun to try to confuse ID apps! I took a photo of this old puffball mushroom and uploaded it to iNaturalist. Four of the five suggested species weren't fungi at all, but *snails.* The last possibility, the pear-shaped puffball (*Apioperdon pyriforme*), is a pretty decent match, but it wasn't at the top of the app's list.

Most apps rely only on algorithms for identification. My personal favorite is iNaturalist because, in addition to the initial algorithmic results, you can get additional help verifying an ID from other iNaturalist users. Many biologists, mycologists, birders, herp enthusiasts, and other naturalists use this app regularly, and it's very common for someone else to either agree with your ID or suggest something different within a day or two of you uploading your observation. You can also leave a comment or question for them in the notes of your observation (make sure you type their username with an @ in front of it to tag them). While you always want to double-check the species being suggested by other users, in my years of using the app, I've had only a handful of instances where they were obviously wrong.

When deciding what apps you might want to download, here are a few things to consider:

What are you hoping to identify? Some apps, like Blossom or Picture Bird, focus on specific groups of organisms, while others, like iNaturalist and its more kid-friendly app Seek, are more general.

What do you want the app to do? Most ID apps make suggestions based on algorithms, but the Audubon Bird Guide is an example of an app that's essentially a digital field guide that updates every time the app does.

What do the reviews say about the app? If it gets consistently good reviews, it's still not a bad idea to check the one- and two-star reviews to see if there are any features people regularly complain about.

Who else uses the app? Certain apps like iNaturalist and eBird are frequently used by professionals and laypeople alike for identification out in the field. Incidentally, both of those apps are scientifically rigorous enough that scientists use the observations collected with them for research purposes.

How much room do you have on your phone? Some apps take up a lot of space, and a phone with little free memory can slow down significantly.

How much time and attention do you have to dedicate to learning and using ID apps? Each one is set up differently and takes time to learn and assess.

While I've tried numerous apps, I keep coming back to iNaturalist because it's all-encompassing, it has the additional layer of human verification, and I'm very familiar with it after using it regularly for several years. That said, as with any other tools, I encourage you to explore various apps when and as you have the time, and see which ones work best for you.

Other Naturalists

Sometimes you just want to talk to a known expert directly about an unusual specimen you've found or one that you're having trouble identifying. While online groups can be a great resource, the quality of the answers you get may be variable, and the algorithmic quirks of some sites like Facebook may mean your post gets buried without anyone else ever seeing it. So, who you gonna call? Here are some possibilities:

- Fish and wildlife departments, state and national parks, wildlife refuges, and other public land management entities often have information on native species on their websites. You can also contact them with questions about species you've observed within their jurisdictions.

- Nature-focused nonprofit organizations such as your local Bird Alliance/Audubon Society or your state's native plant society can often also answer questions about relevant life-forms. Nonprofits involved in advocacy and outreach, habitat restoration, and other conservation efforts also often have staff or board members who are naturalists of one sort or another.

- College and university biology, ecology, and other natural science departments may have faculty or grad students who can answer your questions or help with identification.

- Guides and other people involved in ecotourism are often a wealth of natural history knowledge. This is especially helpful if you're traveling someplace where you aren't familiar with the local biodiversity. Plan to take a tour or hire a private guide to help you learn some of the local species you find along the way.

I recommend sending an email rather than calling whenever possible. First, it allows people to get to your query at their own pace, rather than interrupting them during their workday. Second, an email allows you to send them pictures or videos if you have them, and they may respond with more pictures or other informational attachments in turn.

You may also have opportunities to get out in the field with other people. Birding groups, native plant societies, mycology groups, and other such organizations of nature enthusiasts often have group outings, although some may be for members only. They're a great opportunity to meet other people interested in local nature, who may very well be able to help you identify the species you're stuck on. Plus, those outings are excellent learning experiences, and the more people are present, the more eyes there are to notice cool animals, plants, or fungi!

Do be aware that humans are prone to error, even those with a lot of experience. (That includes me, by the way.) This is where "trust but verify" is a really useful maxim to follow. If something someone says doesn't match up with what I've read in other sources, I automatically make a mental note to double-check that information later. Sometimes they're bringing up more recent information that hasn't had a chance to disseminate as far as previous assumptions, and I might find sources that corroborate what they have to say. Other times? They're just plain wrong, at least on that one detail.

Other Helpful Identification Tools

Everyone's field kit is a little different, depending on where they want to go to look for things to identify. I prefer to travel lightly, so my daypack generally has one set of binoculars and a small magnifying viewer in which I can put insects and other small specimens. If I'm going out to survey birds or other wildlife in a large open area where I will be mostly stationary or driving from place to place, I may bring a spotting scope on a tripod, but I find this to be too large and unwieldy for longer hikes. That said, I'd like to discuss these tools and a few others you may find useful in your identification adventures.

Binoculars

These are the most common tool for viewing birds and other wildlife. They come in a variety of sizes and powers, but the first thing you'll want to pay attention to are the two numbers (aka the "specs") that are usually printed on the center hinge of the binoculars. The first number refers to the magnification power of the lenses; an 8× set will magnify things eight times larger, while a 10× magnifies ten times. The second number is the diameter in millimeters of the objective lens, the larger lens that you point at whatever you're looking at. The wider the objective lenses, the larger your field of vision (the area you can see through the binoculars), and the easier it is to keep whatever you're looking at in view.

I often suggest 8 × 42 binoculars as a good starting size. Unless you are out in a very large grassland, wetland, or other open area, 8× is sufficient to magnify most of what you'll see. If you think you'll be spending more time in those open areas, you might want to get a set of binoculars with higher magnification or a spotting scope (more on that in a moment). There is such a thing as too much magnification. If you are looking for birds in a forested area and see one thirty feet away, if you have a 10× or 12× set of binoculars, they might magnify the bird so much that it won't all fit in your field of vision.

Lenses that measure 42 mm are large enough to give you a decent-sized field of vision without being too heavy. Some people prefer binoculars with a smaller objective lens, such as 28 mm, because they weigh less and are easier to fit in a backpack. Other people decide they want even larger lenses to see more through them and are willing to deal with lugging around the extra weight.

If you aren't entirely sure what you want in a set of binoculars, plenty of brick-and-mortar sporting goods stores, nature centers, and other businesses sell binoculars that you can try out before you make a purchase. Their employees should also be able to give you some advice on what set might work best for you. If there's a particular set I'm looking at, I might check reviews for them on Amazon, especially the one- and two-star reviews. I'm specifically looking for problems that people consistently complain about; for example, if many reviews mention that the insides of the lenses tend to fog up in humidity, that means the seals are probably bad.

You also don't need to buy the most expensive set out there, nor do you have to buy them brand-new. The binoculars that ride around in my backpack set me back about fifty dollars; they work just fine and I'm not out much money if I manage to break them. On the other hand, if you want to invest in the really, really nice Swarovski set you saw at the nature center because they're from a reliable manufacturer, treat yourself!

To use binoculars, look with your naked eyes, and then when you see something of interest, keep your eyes on it while bringing the binoculars up to your face. You should now be able to see whatever it is more clearly, unless you moved your head down while bringing up the binocs, or you were watching an animal that suddenly darted out of the way.

Spotting Scope

Most folks can get by with a pair or two of binoculars. However, if you want a really powerful tool for viewing wildlife and other nature at a distance, consider a spotting scope. This will generally have one set of lenses instead of two, but the magnification will be much stronger and often adjustable. My spotting scope adjusts from 25× to 75×, which makes it really handy if I'm stomping around some wetlands where I may be trying to see things that are a couple hundred yards or more away. Scopes also tend to have larger objective lenses; mine is 70 mm, but bigger ones are available.

The trade-off to all that extra power, of course, is weight. Scopes are best used with a tripod because they're difficult to keep balanced with just your hands. It also makes it easier if you're switching back and forth between binoculars and the scope; I might use the binoculars to identify an animal I want to look at, and then step over to the scope for a closer look. Because the scope does have such a large field of vision, I may use it to scan a large area where I might miss faraway animals with my naked eyes, especially if I'm doing so systematically for a survey or similar activity.

The most powerful spotting scopes tend to be on the pricey side, especially if you're buying from a well-known manufacturer like Vortex. That said, plenty of budget options are available, and while those may be fewer in number than the higher-end ones, they're enough to get you started. Plus, if you hang around the right people long enough, you may find someone getting rid of old equipment for cheap; a few years ago, I lucked out and scored my scope for $100 in a silent auction at a birding festival.

Magnifying Tools

There are several types of magnifiers available to help you take a closer look at what you find in nature. This is especially crucial if you're examining some minuscule organism to see tiny details like colors or shapes, or if you're like me and your eyes just aren't very good anymore, so a little extra visual power helps. A magnifying glass is the one that people are most familiar with—a single lens with a handle—and it is usually the simplest and least expensive option. They're widely available at many toy stores, nature centers, and museum gift shops.

A hand lens or loupe is more compact, and it typically has anywhere from one to three lenses. More lenses mean clearer views and less distortion, but humidity can build up between them and cause them to fog up. Hand lenses with 5× to 20× magnification are common, although there are 30× sets that are more often used by lapidary professionals and jewelers.

One of my favorite tools is a neat little thing sometimes referred to as a bug viewer or magnifying viewer. It's made of two pieces; the base has a small mirror at the bottom and a magnifying lens on the side. A clear cover fits over the base and has another lens on top. If there's something I want to look at but don't want to touch with my bare hands, I can herd it into the cover, then close the opening with the base. There are other styles, but they all allow you to put something small like an insect, leaf, or small mushroom inside and then look at it more closely through the lenses. I keep one in my backpack, especially for classes and other group educational outings where we might all want to get a closer look at something neat we've found. Since some insects and arthropods can bite, sting, or be easily crushed, I prefer this method of capture to minimize direct handling. Amphibians also should not be handled directly, as we often have salts and oils on our hands that they can absorb through their permeable skin, and those can make them quite sick. I'll cover handling wildlife a little later in this chapter.

Microscope

You're not likely to bring a microscope out into the field with you, but it's a nice thing to have at home if you want to look at specimens up close and personal. The microscopes most people are familiar with are optical, using lenses to magnify objects. A simple microscope has a single lens, while a compound microscope has multiple lenses and is more powerful. With a traditional microscope, you look through the lens (eyepiece) at the top of the tube and use the coarse and fine adjustment knobs to bring what you're looking at into focus. The stage is the platform that holds the slide or whatever else you are looking at, with a light beneath it to illuminate it.

Some microscopes have multiple objective lenses—the lens pointed at whatever you're observing—that allow you to change the magnification power; these can be rotated in a circle until the one you want is in line with the tube and eyepiece. Many microscopes meant for general consumer use come with a few objective lenses in commonly used powers. You won't be able to see atoms with them, but you might get a good look at some plant cells or bacteria.

Digital microscopes project the image onto a screen instead of having you look through the eyepiece, although some allow you to do both. This can be easier for people who find focusing through the eyepiece lens a challenge. Some digital microscopes even work with your phone's camera and have camera mounts that allow you to take pictures through the eyepiece of a traditional microscope.

Measuring Tape

It's not a bad idea to have a measuring tape with you in the field. The flexible sort used in sewing rolls up easily into a compact form and is lighter than the stiff metal tape measures with hard plastic or metal cases found at the hardware store. They typically come in either 60-inch (5 feet) or 120-inch (10 feet) lengths; the 60-inch length will suffice for, say, mushrooms, but if you intend to seek out the trunks of big trees, you may want the 120-inch length (and a friend to help measure them).

Even if all you have access to is a ruler, that's fine! A small aluminum ruler packed into your kit is lightweight and durable, although make sure the sharp corners don't dig into the fabric and tear it. Some of the field guides I have from Timber Press have a ruler printed on the back cover. And one of my favorite mushroom hunting hoodies that I got from Camp Mustelid has a ruler printed on the left sleeve, just in case I want to see if the giant bolete I found breaks my personal record or not.

Cameras and Recorders

These days, smartphones have taken over many of the roles of traditional tools, whether that's field guides, cameras, or sound recorders, although some people prefer their old standbys. That new iPhone camera might be nice, but to a die-hard film or

DSLR photographer, it lacks many useful functions and range. In the same vein, newer smartphones often have admirable sound recording capabilities, but for high-quality results, you're going to do better with an analog or digital field recording setup.

It's a good idea to keep your smartphone handy either way. If your phone has a waterproof case, it may be a better option if you find yourself in a downpour where more sensitive (and expensive) electronics may be at risk. And should you find yourself suddenly witnessing a rare animal that could speed away at any moment (was that an ivory-billed woodpecker?!), you're going to be able to pull out your phone and snap a picture or take a video a lot faster than with most cameras.

Journals and Other Records

If you want to record the various organisms you find and identify, you have several options. Many naturalists like to use a journal of some sort to write down notes about what they saw, when, and potential identifications if it's a new species. You don't have to have anything fancy; a notebook or simple blank book will work just fine. Those of an artistic bent may enjoy using colored pencils or paints to illustrate their notes. You don't have to be a professional scientific illustrator; I've sketched pertinent details of a species using a ballpoint pen, with results that my five-year-old self would have scoffed at. If you don't feel like writing or drawing, a voice recording works, too; you can bring along a standalone recorder or download a phone app.

Some apps, like iNaturalist, eBird, and Merlin, will save your observations so you can go back and look at them later. For birds, in particular, you can also download and print out life lists, which are checklists of all the species in a given area, such as the United States, North America, or (for the really ambitious) the entire world. As you observe each species for the first time, you check it off and include information like when and where the sighting happened. I have something similar to a life list; I keep one

Your journal, whether traditional or digital, is a great opportunity to record information about your observations. Be as detailed as possible when describing a species you've seen, and use your own words if you don't know technical terminology. "A slender snake basking on the trail, about fourteen inches long, with a small, oval-shaped head, tapered tail, and dark green, almost black coloration with very pale, thin yellow stripes running down its back and sides, which slithered away quickly as I approached" is much more thorough than "A green and yellow-striped snake."

particular bird field guide in the door pocket of my car, and if I see (or hear) a new species that I can positively identify, I check it off in its entry in the book, along with any pertinent information.

You likely won't need every single item listed in this chapter, which would make for a very heavy backpack indeed! Consider this an inventory of helpful tools to consult before you wander off into forests and fields, and choose whatever equipment will best suit your needs (and budget). In the next chapter, I'll discuss what to do once you're actually out in the field.

CHAPTER 4

Heading into the Field

Joshua tree (*Yucca brevifolia*)

YOU NOW HAVE A BETTER IDEA OF SOME OF THE identification tools that are available to you, and, assuming you haven't keeled over from sheer information overload, you might want to know where to begin looking for cool things to identify. The great news is that pretty much anywhere outdoors is a great starting point! As much as we humans in the Western industrialized world have managed to convince ourselves that we are separate from the rest of nature, we remain surrounded by and immersed in it.

If I go for a walk in downtown Portland in the middle of the day, I'm certainly going see many of my fellow humans and our creations. But I'll also see some feral pigeons (*Columba livia domestica*), which, as domesticated animals, are also a human creation of sorts derived from wild rock doves. And I'm likely to see—or at least hear—a raucous band of American crows (*Corvus brachyrhynchos*). These birds may have to scatter if a peregrine falcon (*Falco peregrinus*) or red-tailed hawk (*Buteo jamaicensis*) decides to look for a snack. I'll no doubt see plenty of trees like Douglas fir (*Pseudotsuga menziesii*) and various maples (*Acer* spp.), along with a whole host of smaller plants like Oregon-grape (*Berberis aquifolium*), licorice fern (*Polypodium glycyrrhiza*), and even some invasive weeds like common chickweed (*Stellaria media*) and herb-Robert (*Geranium robertianium*). Believe it or not, there are even fungi in this heavily developed area, such as splitgill mushrooms (*Schizophyllum commune*) and some ink caps and other *Psathyrellaceae* species.

My point is that you don't have to go very far to find at least some species to identify. I will, of course, find greater biodiversity in wilder, less developed areas further away from town, but I am the sort of naturalist who is constantly watching for other species anywhere I go. Every discovery is a delight, from an unusual slime mold on a log in the Columbia River Gorge to one of the many spiders that occupy windowsills and quiet corners in my home while being on "bug duty."

So, assuming your neighborhood is safe to walk around or otherwise explore, go ahead and start there! You might be surprised by just how many species you find on a relatively short jaunt around the block. Plus, there's something really cool about becoming a bit of an expert on your nonhuman neighbors. I find that knowing who's who in local species fosters a greater sense of connection to any place I go, not least of which is my own home turf.

If you want to expand your range further but don't want to head out into the wilderness, look for local parks and other natural areas that have sidewalks or well-maintained trail systems. Wilder, less developed parks will tend to have greater biodiversity, but don't turn your nose up at someplace with a grassy lawn.

Many naturalists relish the opportunity to visit more rural areas, especially habitats that still have relatively intact native ecosystems. This often includes hiking trails, primitive campgrounds, and off-trail land. These places tend to offer a greater variety of species to identify and to include many that may not be able to thrive in cities or other developed areas.

Here are some tips to get the most out of your outdoor identification excursions:

Be prepared: Dress for the weather, even if you're just heading out into your own neighborhood, and make sure your footwear is sufficient for the terrain. If you're going further afield, take a daypack with essentials like a first aid kit, water, food, and so on (you can read more about these by searching for "ten essentials" at nps.gov). Take a map of the park or trail system; some areas may have a map near the parking lot you can take a photo of. Should you be heading off-trail, make sure you have both a GPS device and a topographic map and compass as backup—and know how to use them.

Take your time: Many walkers and hikers want to cover as many miles as they can in a day. However, it's common for us naturalists to be the exact opposite— we can't go very far without getting distracted by some neat mushroom, plant, or bug. That's okay! Think of it as emphasizing quality over quantity. Even if you make it only a couple of miles—or a couple of hundred yards—you haven't wasted your time.

Be nice to the land: In many developed parks, you're encouraged to wander across the grass instead of sticking to the sidewalks. That's usually not the case in more dedicated natural areas, where the focus is on preserving as much biodiversity in the ecosystem as possible. If there are established trails, stay on them; they're there to take the brunt of the foot traffic so the rest of the place can remain untrampled. Should you venture into off-trail land where cross-country travel is your only option, do your best to leave no trace. Most public lands have rules for allowed uses, and it's best to stick to them. If a particular area is off-limits or specimen collection isn't allowed, respect those boundaries. Avoid trespassing on private land; in the event a private landowner gives you permission to visit their land, again, practice leave-no-trace ethics.

Accessibility and Nature Identification

Let me preface this section by saying that, beyond needing glasses, I don't have any physical disabilities. I probably have some combination of various undiagnosed neurodiversities, but I've found many workarounds over the years (like combining hyperfocus and infodumping to channel a whole bunch of information about my biggest special interest—nature—into this book).

Throughout the next few chapters, I focus on identifying different groups of organisms through a variety of physical senses, particularly sight, hearing, and touch. These are senses that I confess I take for granted; I know that some people are limited in the ways in which they are able to practice observation and identification, whether that means being unable to see or hear, or having restricted mobility that prevents them from venturing beyond their yard or going outside at all. I know brilliant naturalists who have trouble reading because of dyslexia or attention issues and who may not be able to spend hours poring over field guides and other books.

Unfortunately, accessibility accommodations are often subpar to nonexistent, and that includes the outdoors. There are scant few tools that may help people with some disabilities in their nature observations. Digital binoculars with significant zoom capabilities may be of use to some people with visual disabilities, while someone who is hard of hearing or deaf could use the Merlin app to identify bird calls not just through sound but also via live spectrograms. Both Syren Nagakyrie's *The Disabled Hiker's Guide* series and *The Feminist Bird Club's Birding for a Better World* by Sydney Golden Anderson and Molly Adams occupy an all-too-small niche of books on how to make the outdoors more accessible.

You may have luck reaching out to other people who have similar disabilities or conditions, as they may have suggestions on what worked for them. Organizations like Disabled Hikers (DisabledHikers.com), Birdability (Birdability.org), Accessible Nature

(AccessibleNature.info), and Access Birding (AccessBirding.com) may offer resources for accessible outdoor activities, networking with others to create opportunities to get outside, and educating public officials, businesses, and other entities about the dire need for better accessibility in outdoor spaces.

One of the benefits of going outdoors with other people is mutual support. A blind person may not be able to see a bird nearby, but they may be the best in the group at birding by ear. Another person could describe the bird's colors, shape, and other traits or page through a field guide for information. It's common for hearing people to lose their ability to hear higher registers as they get older, meaning that many older birders can no longer hear the calls of kinglets or juncos, but they could point out one of the little birds hiding amid the underbrush.

The nice thing about the growing wealth of identification resources out there is that you can choose the ones that work best for you. When I say, "Use all of the resources you possibly can," that means work within your realistic parameters. For example, if you can see but have trouble looking at a screen for more than a few minutes at a time, don't worry so much about websites and other online sources. Try focusing on physical field guides, connecting with other people in person or on the phone, and other analog solutions. With time you'll find what combination of tools and materials are most effective for you, and I hope there will be some ideas in this book you can incorporate into your identification practice.

My Identification Process

So, what does your beloved author do when she finds a mystery species? The following is my typical order of operations, although I don't always stick strictly to it in the form presented. I'm sharing it mainly to give you some idea of my thought processes behind each step, as well as one possible way to go about identifying something if you have no idea where to start.

Observe

When I see an unknown species, I do my best to notice as many details about it as I possibly can. The next few chapters go into greater depth on what to look for in each group of beings, but some common traits include colors and patterns, size, overall shape, detail shapes (wings, beaks, leaves, caps, and so on), location and habitat, and, in the case of animals, behaviors and movement. If I'm able to get some good pictures, that helps a great deal. Plants and fungi tend to be quite accommodating for even the most amateur photographer, and this is when I might decide to take a small specimen. Animals are tougher since I'm usually carrying just a phone camera and often can't get a good shot before they move away. In that case, I jot down as many traits as I can, either typing them in as a note on my phone or sketching them on a scrap of paper.

Consult Apps

If I've been able to get at least one decent photo (or sound recording), my next step is to put it on iNaturalist. It usually offers me a few initial species suggestions to look up, unless I've managed to find something that completely confuses it (like LBJs, LBMs, or DYCs; see the sidebar). When I'm in a hurry, I'll either pick the species that seems like the best fit or go up the taxonomic ladder until I get to a level I feel comfortable choosing; it's not uncommon for me to have initial observations labeled "Poaceae (grasses)" or "Fungi including lichens" as placeholders. Should I have more time, I'll go through iNaturalist's suggested species, look them up online, check any field guides I have with me, and otherwise do a more detailed assessment. If I'm still not feeling confident in any of them, I'll jot down a placeholder and do more research at home.

LBJ = "little brown job," aka "little brown bird"; any of a number of similar small passerine birds that includes several species of New World sparrow.

LBM = "little brown mushroom"; any of a number of small, nondescript brown mushrooms that are tough even for experienced mycologists to differentiate.

DYC = "damned yellow composite"; any of a number of Asteraceae species whose yellow composite flowers look extremely similar to one another. For those unaware, the blooms of the Asteraceae, including daisies, sunflowers, and many more, are not single flowers but are instead clusters of many tiny disc florets in the center, surrounded by petal-shaped ray florets.

The other advantage to starting with iNaturalist is that it makes my observations available to other app users that much more quickly. If I'm spending several hours on a trail and then driving home, I might have a whole string of good suggestions to follow up on once I get home. I still give these crowdsourced suggestions the same research treatment as the algorithmic ones, rather than simply accepting them as correct.

In the specific case of bird calls, I immediately open Merlin for a live listen. The scrolling spectrogram displays a visual representation of each sound the phone picks up (which is why it's important to stay quiet during recording). Below it, Merlin lists the birds it recognizes, and each one's name lights up when it sings or calls. The app saves the recording once I'm done, so later I'll open AllAboutBirds.org and listen to a few sound files of each species that Merlin identified and make sure they match what I heard.

Do Research at Home

Once I've made it back home, I start pulling out the relevant field guides. If I have some good ideas, whether from my own observations and experience or from app suggestions, I'll generally start with those species to see if, in fact, they do match whatever I saw. Should none of those pan out, I need to identify other potential options. In some cases, I can pretty confidently say I saw a member of a particular family or genus, and so focusing on those groups in my field guides often yields a good species match. The table of contents of most field guides break their contents down by family, but if a book has a different sort of arrangement, you can also often find families and genera listed in the index.

I also do some online searches for the species in question. I'm primarily looking for photos to verify that this is, in fact, the organism I observed. However, I also want to confirm the species' range, find out about any similar species, and see if there have been any taxonomic updates like new names or species being combined or split apart. In a few cases, I've learned that the species I observed is scarce in my region and worth keeping an eye out for.

There have also been plenty of times when I've been completely stumped, with no idea of where to start. In that case, I'll just start paging through field guides and looking at pictures, putting a bookmark at every page that has a potential match. Then I research the species in more detail, using books, websites, search engines, and even other people's observations of the species.

Some apps, such as iNaturalist and eBird, allow users to see one another's observations. This might consist of a map or a series of lists, and they're a great way to see what other people have been finding. If you're able to search by species, you can confirm whether the one you think you've found in a particular area has been observed there before. Because new observations are always being added, these app records are often more up-to-date than the range maps in field guides.

Ask for Help

Sometimes, even after all of that work, I still might not have a good answer for what I've found, or I want some external confirmation of my identification. That's the point at which I start contacting other people. I know many naturalists personally, so I'll usually talk to them first. I may also email area biologists and other scientists to ask them what they think about what I found, as well as post to relevant online groups asking for their consensus on what I saw. Once again, if anyone recommends a species to me other than what I thought I had, I'm going to do my due diligence and run that species through the above process, too.

Even after all this effort, it is entirely possible that I still don't have a firm identification, at least down to the species level. And you know what? That's okay! I just identify as far down the taxonomic ladder as I feel confident, and then I stick a pin in it. Maybe later I'll get more information that verifies the species, or someone will pop up on my iNaturalist observation with a new suggested taxon. I do the best I can with what I have at any given time, and as an everyday naturalist, it does me no good to be so focused on identifying every being down to the species level that my nature identification becomes a lot more obsessive and less fun.

Start with Something Easy

I've given you a lot of information in this chapter, so it's understandable if you're feeling overwhelmed. If you're not ready to start identifying every single living being in sight, pick something easy to start with. Here are a few suggestions:

Practice identifying a species you already know:
By practicing on a familiar species, you're getting a chance to get used to the process of identification and the tools and skills involved therein. You may already know how to pick out a common dandelion

(*Taraxacum officinale*) from the crowd, but how might you figure out what it is if you've never encountered one before? Or, for that matter, if someone asked you how you know this is a dandelion (and not another species of DYC), what would you tell them?

Try something that you can already narrow down to the order or family: Let's say there's a mushroom that you're pretty sure is in the family Boletaceae because it has a convex cap with spongy pores on the underside supported by a clearly defined stipe, and it is growing out of the soil. If you focus your research on that family, there's a good chance that you'll find out which species it is more quickly than if you were trying to identify a mushroom where the best you could say is, "Well, uh, it's a fungus."

Pick something that seems to be common: If you keep seeing large numbers of small birds of the same type, it's probably easier to figure out what they are compared to a scarcer bird. Not only are they more likely to show up in your field guides and other resources, but chances are other people see them regularly too and can tell you what they are. Moreover, the more common a species is, the more time you can observe it and take note of its traits, behaviors, and so on.

Finally, I again want to remind you: Every single animal, plant, fungus, or other living being you meet is an opportunity to practice nature identification. The more you practice, the better you get!

Specimen Collection

Let's talk about bringing things from nature back home for further examination. This has been a longtime practice of naturalists, and it has offered many opportunities to discover details about a species that aren't readily noticeable out in the field. At the same time, it's an inherently extractive practice that removes resources from ecosystems, often permanently. While it may not seem like a big deal to take a few leaves or mushrooms, in an era when habitat loss is rampant and popular public trails and parks can be ravaged by many people "just taking a few flowers," it's important for us to consider our individual and cumulative impact.

First, ask yourself whether you really need to take a specimen home. If it is an animal, the answer is almost always going to be no. Every animal you remove from an ecosystem is one less set of genes adding to the diversity of its local population. Moreover, as a general rule, wildlife do not make good pets. Captive wild animals often have very specific care needs that most folks are not able to provide, and they may not have as high a quality of life as their wild counterparts. Keeping an animal captive for even a few days can cause them to become quite stressed, which can sicken or kill them. I extend this not just to vertebrates, but invertebrates as well; I would rather that the neat bug I found be able to live out its life outside where it may get the chance to procreate and probably become food for something else, rather than be hidden away for years dried and pinned to a piece of Styrofoam.

There are cases where the only way you can positively identify something down to the species level is for an expert to examine it in the lab, which, of course, involves removing it from its habitat and sending it in for dissection. This is especially common with insects and other arthropods. However, I am not involved in field research, habitat restoration, agriculture, or other professions where specimen collection for identification may be crucial, and I personally don't feel that it is important enough for me to know the exact species of a little critter like that for me to send it off to its death. If I can't get enough information from my observations and photos, I'm content identifying it as far down the line as I can and then leaving the species as "unknown" or "maybe this, maybe that."

There was a sad trend during the latter part of the nineteenth and early twentieth centuries when a significant number of North American animal extinctions occurred. If it became evident that a species was becoming scarce, zoos sought out live individuals to display (but not necessarily to breed), and museums would send hunters out to shoot a few of the ones that remained so that they could be preserved as taxidermy or other specimens. Today, many rare species are further diminished by illegal demand from private individuals. It's a big part of why my personal specimen collection ethic has shifted in recent years toward "Take only what you need and leave the rest." If you're unsure of how much is needed, err on the side of the bare minimum— a leaf or two, for example, not an entire branch.

What about handling a live animal in the field? Some people think we should never touch any wildlife, while others believe it can be useful in limited cases for education and identification. Most of the time when naturalists catch and examine wildlife, it's for scientific purposes, such as surveying and tagging birds. But it's not necessary for an everyday naturalist to spend excessive time handling a wild animal.

Some specimens are illegal to take from the wild. Here in the States, individual states have laws prohibiting the possession of some or all native wild animals, as well as lists of animals and plants considered to be endangered or threatened. Federal laws add further protection for scarce or sensitive species. Many of these laws make possessing either live or dead specimens illegal.

Some public lands also have laws governing them; National Wildlife Refuges, for example, prohibit taking any natural specimens or cultural artifacts from refuge lands, even if they are otherwise legal to possess and collect.

In some cases, it's also illegal to even pick up animals; the Migratory Bird Treaty Act (MBTA) makes it a federal offense to pursue or capture most wild native birds in the US; the Migratory Birds Convention Act (MBCA) does the same for Canada. Handling wildlife, even for a very short time, can really stress them out, so the less we do it, the better. And many animals are downright unsafe to even approach, let alone touch—just look at everyone who's ended up injured when they tried to pet a bison (*Bison bison*) at Yellowstone National Park.

The MBTA and MBCA also prohibit the possession of most wild bird feathers in the US and Canada. During the Victorian era, feathered hats became quite popular, and women would walk around with feathers, dried wings, or even entire taxidermy birds on their heads. Plume hunters went into woods, fields, and wetlands and wiped out entire populations of multiple species, to the point where several became close to extinction.

Because there is no way to tell the difference between a naturally molted feather and one that was plucked from a poached bird, these laws prohibit the possession of live specimens, individual body parts, eggs, and nests of all native wild birds in the States and Canada, except for a few game birds like grouse and turkeys. So, if you find a feather on the ground and you can't say for sure what it's from, take a few pictures and leave it where you found it.

I might occasionally tuck an insect, spider, or small reptile or amphibian into a magnifying viewer for a few moments to let people get a closer look, or briefly catch a garter snake to show a group I'm with that they are not, in fact, slimy or scary.* But for the most part I appreciate the animals from a respectful distance, and I use a lot of discretion to decide whether the impact on the animal is going to be worth the educational experience for others. Any handling for examination or educational purposes needs to be brief, with the animal's safety as the main priority, and they must always be released exactly where they were found. And if you choose to just never pick up a wild animal, that's the surest way to ensure their health and security.

With regard to plants and fungi, it's often easier to take a small sample, like a few leaves or a single mushroom. I may do this with a species that seems to be common in an area, but if I am finding only a few individuals, I'm more likely to leave them be, especially if I don't know what they are. I don't want to run the risk of damaging an endangered, threatened, or otherwise scarce species. In those cases, I simply take many photos from all angles and of varying details as well as the whole organism.

If I am going to take a plant or fungus specimen, I remove as little as I need to help get my identification and always with an eye toward minimizing damage. For example, if I want to collect a small lichen that grows on trees, I look for one attached to a branch or a piece of bark that has already fallen off. If that's not an option, I carefully excise one lichen from its home, or a fragment if possible, and again take photos of the entire colony. It's the same practice with any plant or fungus specimen—I take the minimum amount I need for my purpose and no more.

I am not so cautious when it comes to very common species; here in the Columbia-Pacific region, Sitka spruce (*Picea sitchensis*),

---

* A study in the journal *Biological Conservation* in February 2024 showed that positive experiences with snakes lead to people having improved attitudes toward them. If holding a garter snake for a minute or two means that the group I show it to is less likely to harm or kill snakes in the future, I figure it's worth it (and to get garter snake musk on my hands!).

Many lichens grow very, very slowly. Some species grow only a fraction of a millimeter a year on average, and so damage to a lichen or a colony thereof may take many years to repair. It's not just lichens that are a concern, either. In her book *Gathering Moss: A Natural and Cultural History of Mosses,* Robin Wall Kimmerer explains how huge sheets of moss torn off trees to be sold at craft and garden stores may never grow back at all.

salal (*Gaultheria shallon*), and turkey tail (*Trametes versicolor*) are quite abundant in spite of widespread clear-cutting in the Oregon Coast Range and Willapa Hills. If I want to grab a few specimens to draw in a sketchbook—along with some fresh spring spruce needles for tea—I don't feel bad about it. And as far as invasive and other nonnative species go, my philosophy is "The more removed, the better."

You aren't necessarily going to know if a given species is rare or common or native or invasive if you haven't yet positively identified it. So, until you know what you're dealing with, it's best to err on the side of caution and leave it where you found it.

For the most part, when I am done with a leaf, mushroom, or other specimen, I put it back where I found it, or in the ecosystems near my home (assuming there's no chance of spreading an invasive species). This allows its nutrients to be returned to the food web and helps me keep clutter in my home at least somewhat under control. If you have a good reason for hanging onto a specimen, though, you'll want to research ways to preserve it long-term.

There's not enough room here to go into all the methods of specimen preservation, but the internet is a great place to start looking for options for the type of specimen you have in mind.

Some are relatively simple; bones and (legal) feathers just need to be dusted off now and then—and watch feathers for any signs of moth or other insect damage. Flowers and other herbaceous plant parts are most often pressed and dried, although some can be preserved in a more three-dimensional form with glycerin. And some preservation methods, such as hide tanning or taxidermy, can require significant time and resources and may be challenging for a complete beginner.

Now that you have a better idea of how to practice nature identification in general, it's time to explore some specific considerations for individual kingdoms, starting with the animals!

CHAPTER 5

How to Identify Animals

snowflake moray eel (*Echidna nebulosa*),
red cushion sea star (*Oreaster reticulatus*)

AS A NATURE TOUR GUIDE, I TRY TO TAILOR EACH
tour to the participants' interests. Some may be avid birders, while
others are interested in learning as many edible wild plants and
fungi as possible. Regardless of what the tour's focus is, I also try
to couch that information within the context of the entire ecosystem. After all, species do not live in isolation from one another,
but as a vibrant, interconnected community made of countless
relationships and dependencies.

I meet many people whose idea of "seeing cool things in
nature" is primarily limited to wild animals. That's not at all
surprising; wildlife viewing is a multitrillion-dollar global industry.
While there are certainly folks who enjoy traveling to see new
and interesting plants or fungi, the animals tend to get the most
attention overall.

Animals are also often the most easily recognized beings
in nature. Your average elementary school kid in the United
States might not be able to identify an acacia tree (*Acacia* sp.) in
a picture of the African savanna, but they can easily identify the
lion (*Panthera leo*) resting in its shade. That doesn't change much

for many people once they're adults. Even among nature lovers, it's common for people to ooh and aah as a bird flies overhead for a few seconds while walking past a plethora of plants with nary a glance.

On the bright side, our focus on animals means that they're a great gateway to the rest of nature; once you start digging deeper into an animal's natural history, you find out more about the other species they are interdependent with, both animal and otherwise. And they're pretty fascinating beings in and of themselves, too. Even if you ever identified only animals, you could spend your entire life discovering new species and never learn them all.

Basic Animal Anatomy

Try to describe what an animal typically looks like. It has four limbs, right? Well, only if you're a tetrapod, which certainly doesn't include centipedes, hydras, or squid. What about "All animals have heads"? Oysters don't even have central nervous systems, let alone heads. And if you guessed, "Animals are living beings that can freely move around," well, you're forgetting about sea anemones and sponges, which spend most of their lives anchored in one place after a brief childhood floating in a sea of plankton. Unless you're willing to get down to the cellular level, it's tough to find anything that can be used to describe all animals, beyond "a life-form organized around a tube-shaped digestive system."

That's what makes trying to describe basic animal anatomy rather difficult, compared to the brief anatomical sections in the plant and fungus identification chapters. From sponges to sea slugs, centipedes to pseudoscorpions, and minnows to monkeys, animals cover a wide range of body plans, both internally and externally. The primary division in the animal world is invertebrates versus vertebrates. Invertebrates do not have a spine or any other bones for that matter. Some, like octopi and worms, are soft-bodied, while others, such as crabs and wasps, have hard exoskeletons armoring them.

emerald swallowtail (*Papilio palinurus*),
Asian vine snake (*Ahaetulla prasina*)

Animals with notochords (spinal cords) (Chordata), also known as chordates, usually have skeletons made of calcium, or less often, cartilage. This includes all animals within the subphylum Vertebrata, and the term *vertebrate* is commonly used to refer to all chordates.* However, the subphyla Cephalochordata (lancelets) and Tunicata (tunicates) lack skeletons but do have the notochord and other traits that separate them from invertebrates.

The invertebrates, for their part, are even more diverse. Rather than being a single phylum, *invertebrates* is a paraphyletic category that essentially boils down to "all animals except for the chordates." Their main common feature is their lack of a notochord, and beyond that they have evolved into a dizzying array of forms. There are more than 1.3 million known species of invertebrates, compared with slightly more than 66,000 known vertebrate species, and they are as varied as starfish, coral, and spiders.

Even when you try to categorize animals by something as seemingly universal as our digestive systems, there are exceptions. Within days of an animal egg cell being fertilized, it undergoes a process called gastrulation, in which it becomes a multilayered mass with a single orifice. In deuterostomes, which include chordates and the phyla Hemichordata and Echinodermata, that orifice eventually becomes the anus, and the mouth develops later. In many protostomes, which includes all other bilateral ("two-sided") animals, the first orifice becomes the mouth. But then you have creatures like sponges (Porifera), jellyfish, and their kin (Cnidaria), which are neither deuterostomes nor protostomes but have their own forms of embryonic development.

---

* It's actually a bit stickier than that; some taxonomists propose that the phylum Vertebrata (animals with vertebrae) is a less accurate term than Craniata (animals with skulls). This is because hagfish (Myxini) have skulls, and their notochords are protected by cartilaginous structures similar to vertebrae but considerably more primitive. They are often thought to be closely related to the lampreys (Hyperoartia), which have similar long, slender bodies but with cartilage skeletons. Together, the Myxini and Hyperoartia may be known as the Cyclostomi, but how closely related those two groups are to each other and to the rest of the vertebrates is still under debate.

This isn't even getting into the anatomical terminology associated with specific groups of animals, such as arthropods or echinoderms. As we saw in the section on dichotomous keys in Chapter 3, even some of the words for mammalian body parts can get technical. And since some groups of animals may seem alien in appearance to us, it can be a challenge sorting out which part is which.

Feeling confused? That's okay. When you see an animal you don't know yet, you're still likely to be able to recognize and name at least some of its physical features, like legs, wings, or tentacles; eyes, ears, and mouths; and so forth. While you'll encounter more specialized terms as you get into more detailed anatomical discussions, most of us have learned the basics by the time we hit adulthood. Let's get into some more easily recognized traits to help with identification.

Color

When we describe animals, we often keep the colors simple. "I saw a brown rabbit" or "Did you see that yellow bird over there?" Sometimes that's the best you're going to get if the animal you're observing is in motion or far away. But if you have a chance to get a closer look, you're likely to notice more nuances in those color schemes. An eastern cottontail rabbit (*Sylvilagus floridanus*) has a grizzled mix of tan, chestnut, gray, white, and black hairs over most of its body, with a white belly and underside of the tail. You might notice white rings around the eyes, a white chin, and pale tan on the muzzle, with bright chestnut forelegs. If you saw an adult male of the northern subspecies of the yellow warbler (*Setophaga petechia*) just for a moment, then yes, he would appear as "a yellow bird." But if he hangs around a while, you'll see that his wings and back are green-tinted, and he has reddish-brown streaks on his breast. His beak is dark gray and his legs are peach-colored, with what looks like a solid black eye.

Notice how I am not only listing the individual colors, but also where they are on the animal and what patterns they form? I could have just said the warbler was "yellow, green, and brown," but that's not nearly as descriptive. Especially since similar species may have only minor differences in color, it's important to gather as much detail as you possibly can with regard to hues, shades, patterns, and percentages of the body covered in each one.

One thing that I find helps is sketching out the animal and either coloring it in or labeling different parts with their respective colors. I do this as soon as I can after I've seen the animal, especially if I got to view it for only a few seconds. This helps me to record the colors and patterns while they're still fresh in my head. If I see something at home, I can go inside and grab some markers or colored pencils to do a more detailed drawing, but in a pinch, a single pen or pencil will work. Since I don't have multiple colors, I will sketch in the markings I see, write in all the colors I observed, and then draw lines between the markings and their respective colors.

Don't worry about describing your animal using the "correct" terminology; you don't have to know the difference between remiges and coverts to be able to point out that some of the longer feathers on the warbler's wings had darker greenish stripes on them. You may very well pick up those and other terms over time, which can help you make an even more accurate description of what you saw. But if you use just basic common terminology, most people will understand what you mean, and in some settings, it's better to stick to plainer language that clearly explains things to a general audience.

Size and Proportions

Imagine you're walking down an alley in New York City and you see an animal roughly the size of a guinea pig but with a long, skinny tail run across your path. It's likely the first thing you think is

"Lattice work" in brown

Brown tail

Pale stripe over eye

Short, dark, straight beak

Like oversized sanderling

Buff belly

Gray legs, not yellow

Your sketch doesn't have to be fit to hang in an art gallery. I can draw passably well (and I swear I write legibly when I actually try), but you wouldn't know it from some of my hurried scrawlings on scrap paper, napkins, envelopes, and so on. My very first black-bellied plover (*Pluvialis squatarola*) had its portrait done on the back of a survey sheet during the Pacific Flyway Shorebird Survey one year. It's not pretty, but it was enough to record the appearance of the bird in my spotting scope so I could look it up once I was back in the car.

"That's a *big* rat!" Similarly, if you see a photo online of a frog perched on an American quarter, you might respond by cooing "Oh, look at the adorable frog—it's so *tiny*!" Even if we don't think about how there may be even larger rodents or smaller amphibians, we at least have a basic appreciation for the remarkable dimensions of the animal in front of us.

Field guides often measure the length of an animal in inches or centimeters. Since you aren't usually going to be able to hold up a ruler to a given animal unless it is very slow, very calm, or very dead, you're just going to have to do your best to estimate its size. Some people have a good eye for this, but if you're not one of them, try using other things around the animal to help give you a sense of scale. If a woodpecker lands at your suet feeder, and you know the feeder is a six-inch square, compare that six-inch span to the overall length of the bird. Your estimate is likely to land within the average size range of its species, which should help you when you start poring over field guides.

California ground squirrel
(*Otospermophilus beecheyi*)

How does the animal compare to common objects? Let's say you are swimming in a crystal-clear creek, and a fish swims by you. Is it as long as a ballpoint pen? How about a football? Maybe it's a big one that's about as long as the handle of that broom you use to sweep out the garage. Again, don't worry about getting the length right down to the exact inch; an estimate will work fine.

Some people compare the size of an unknown animal to a known one. If you saw a nutria on the bank of a river and had never seen one before, you might think it looks something like a muskrat, but one the size of a beagle. Birders often use common birds when assessing the size of a mystery species; for example, they consider whether it was closer in size to a sparrow or a jay.

In addition to overall size, pay attention to the animal's proportions as well. Imagine you are looking for starfish (or sea stars, if you prefer) in the shallows of Puget Sound in Washington, and you're lucky enough to find a fat blood star (*Henricia sanguinolenta*) next to its close cousin, the Pacific blood star (*Henricia leviuscula*). They both have five arms and are a vibrant red color, but on closer inspection, the fat blood star has thicker arms that become especially large close to the middle of the starfish (the central disk). The Pacific blood star's legs are much more slender and do not gain much girth closer to the central disk. Even though these are similar species, proportions are the key to telling the difference between them.

Shape

At first, this may seem like an obvious trait to look for. Snakes are long and skinny, beetles are round with hard wing covers, and everyone knows what a fish looks like, right? Except that we're not looking just at that first general impression; we want to get into the nitty-gritty of what makes one animal species' shape unique compared to another's. There are times when two almost identical species can be differentiated only by tiny variances in shape.

You should get an idea of both the *overall shape* of the animal and the *detail shapes*. The overall shape is how the entire animal is formed, and it can be surprisingly challenging to describe it without just saying something like "That bird is shaped like a bird!" Let's say you see an American bullfrog (*Lithobates catesbeianus*) sitting on the ground, and I ask you to describe its overall shape. It would be easy to say, "Well, it's shaped like a frog. You know, frog-shaped? Everyone knows what a *frog* looks like!" But pretend I am an alien from another planet and have never seen a frog, and I want to know what a frog shape really is. So, you might explain that a frog typically has a squat, round, compact shape overall, compared to the long, slender, curving shape of a salamander. You don't have to elaborate beyond a few descriptive terms; the point is to think about what the characteristics of a "frog shape" are, compared to a "bird shape," "slug shape," and so on.

Detail shapes involve how individual parts of the animal's anatomy are formed, like feet, facial features, tails, topline or back, wings, antlers or horns, and so on. Looking back at your bullfrog, you might notice that it has a rather wide, short face with round, protruding eyes and large, tympanic membranes (ears), and while it has long, muscular hind legs and wide, webbed feet, it tends to keep them folded closely to its body at rest. There's no tail, but its backside comes to something of a point. If you really spend time studying your specimen, you may notice a slightly depressed line down its spine, the points of its pelvis sticking up on the lower back, and the slit-shaped nostrils.

If the animal is in motion and you get to see it for only a moment, it may be easier to observe the overall shape and tougher to get the details, so just do your best with what you have. Anything you can note may come in handy when you're trying to determine what species you've observed, whether you recognize them in a photo or other illustration or find a mention of a particularly distinctive feature in one of your books or other resources. Both the bobcat (*Lynx rufus*) and the Canadian lynx (*Lynx canadensis*)

long-billed curlew (*Numenius americanus*),
American bullfrog (*Lithobates catesbeianus*), red slug (*Arion rufus*)

are medium-sized wild felines whose overall shape is compact, lean, and somewhat resembling an oversized domesticated cat (*Felis catus*). Some detail shapes common to both species include long, relatively slender legs; round heads with prominent tufts of fur on their ear tips and cheeks; and short, stubby tails. But lynxes tend to be larger overall and often heavier in build. The lynx's paws are much larger in proportion to the body than the bobcat's, while the bobcat has a slightly longer tail. The lynx also has longer ear and cheek fur tufts.

This approach works for invertebrates as well. A cross orbweaver spider (*Araneus diadematus*) looks quite similar to a cat-faced spider (*Araneus gemmoides*). Both are comparatively large orange-brown spiders with large abdomens and long, angular, slender legs that are well adapted to moving around a web. The cross orbweaver's abdomen is round and smooth on top, while the abdomens of both sexes of the cat-faced spider have two prominent points that look like cat ears—hence their common name.

Just because something is shaped like an animal you're familiar with doesn't mean it is another one of those animals. Glass lizards (*Ophisaurus* spp.), slowworms (*Anguis* spp.), and other legless lizards resemble snakes, but they are an example of convergent evolution where multiple separate lineages evolved the same physical feature—in this case, leglessness—independently. Lampreys (Petromyzontiformes spp.) are long and sinuous like eels, but they are only distantly related to them, being in an entirely different order. So, if you see your first lamprey and think of it only as "eel-shaped," you might be tempted to just assume it's an eel, as might other people you try to describe it to. Some animals are deliberate mimics; for example, several species of spiders have evolved to look so similar to ants that they fool not only the colony they infiltrate, but human observers as well. That arthropod you see may certainly be ant-shaped, but that doesn't mean it's an ant!

This is where you need to look for detailed shapes whenever possible. Lampreys do not have jaws like eels do, but a large, circular mouth full of concentric rows of teeth. They have only a single nostril, and instead of gill slits, they breathe through a set of round pores on the side of the head. Take a closer look at that "ant," and you may notice that what looks like a pair of antennae is actually a pair of legs, indicating that this is, in fact, a spider.

Fur and Feathers, Scales and Chitin

Animals have evolved a wide range of body coverings to protect themselves from the elements and other dangers, as well as attract mates and communicate with one another. Mammals have hair, birds have feathers, reptiles and many fish have scales, and, for the most part, amphibians have no covering on their skins at all. Invertebrates vary from soft-bodied worms and slugs with no protective layers except a bit of slime to the tough armor of arthropods like crabs and insects. These can often be very important traits in figuring out what species you're observing. We've already talked about color; this section is about other physical qualities.

As with the overall shape versus the detail shapes, don't just stop at saying, "It has feathers; therefore, it's a bird." What is the visual texture of those feathers? Are they long and glossy like those of a common raven? Or do they resemble hair or fur, as on a kiwi bird (Apterygidae spp.)? If you find a feather on the ground, does it feel soft or is it stiff and coarse? Going back to what we talked about with detail shapes, does the bird have any particularly remarkable feathers, such as the crest on the head of a northern cardinal (*Cardinalis cardinalis*) or the delicate breeding plumes of a great egret (*Ardea alba*)?

You can do this with pretty much any animal. The exoskeleton on the heads of many male rhinoceros beetles (Dynastinae spp.) grows into horns or other protrusions; these allow them to battle with one another for the attention of a female. Snakes, lizards, and other reptiles often have unique patterns of scales on their mouths or bellies, or the scales themselves have notable bumps or other contours not found in other species. Sharks may appear to be smooth, and if you were to (safely) pet one from the head to the tail, it would feel that way, too. But pet the shark in the other direction and its tiny scales could easily cut your skin.

A few animals deliberately create their own artificial covering. Some ungulates wallow in mud to create an additional barrier against flies, mosquitoes, and other biting insects. Egyptian vultures (*Neophron percnopterus*) also coat themselves in dirt, but they specifically choose soil that stains their feathers red, possibly as part of intraspecies social activity. Among the most elaborate coverings is that of caddisfly (*Trichoptera* spp.) larvae. They start off life in freshwater and weave a protective casing out of silk; many of them incorporate pieces of gravel, twigs, and other hard objects for camouflage and structural integrity. If you see a tiny pile of debris walking along in the shallows of a stream, there's a good chance it's a young caddisfly in its homemade armor.

To recap, here are a few things to look for when examining an animal's outer covering besides the colors:

Visual texture: What does its appearance suggest the animal might feel like if you touched it? This can include descriptors like shiny, dull, rough, smooth, leathery, rubbery, fluffy, bumpy, knobby, velvety, glossy, wet, dry, and so forth.

Tactile texture: If the animal is safe to handle or you can access a loose feather or fur, how does it feel to the touch?

Unique shapes or patterns: This includes those special feathers on birds, scale patterns on reptiles, remarkable tufts of fur on mammals, and so forth. Keep in mind that you're primarily describing the animal's outer layer (skin, exoskeleton, and so on), not what's underneath it.

Bare versus covered: How much of the animal is covered in fur, feathers, and the like, or not?

elk antlers (*Cervus canadensis*), rough-skinned newt (*Taricha granulosa*),
spiky horseshoe crab (*Limulus polyphemus*)

A Carolina mantis (*Stagmomantis carolina*) is encased in a watertight suit of chitin, while a Pacific tree frog's (*Pseudacris regilla*) semipermeable skin must remain uncovered so it can breathe through it. A turkey vulture (*Cathartes aura*) is mostly covered in feathers, except for its head, lower legs, and feet. And the Olympia oyster (*Ostrea lurida*), like other bivalves, can entirely hide itself in its calcium shell by closing it, but it must open it to filter-feed.

Life Stage

When an animal is born, it rarely looks exactly like it will when it grows up. Animals go through several stages of development on their way to maturity. Some, like many species of snakes, have the same basic shape as adults but go through changes in color and pattern as they grow. Insects, on the other hand, go through multiple distinct stages of metamorphosis; while *hemi*metabolous species such as grasshoppers hatch from their eggs looking like miniature adults, *holo*metabolous species may undergo particularly dramatic changes in morphology, such as caterpillars transforming into butterflies or moths or grubs into beetles.

Beware, though—"smaller" does not automatically mean "younger." More than one birder has died a little inside when someone pointed to a fully grown adult downy woodpecker (*Dryobates pubescens*) and said, "Oh, look, it's a cute baby!" The downy is half the size of the hairy woodpecker (*Leuconotopicus villosus*), but their almost identical coloration sometimes leads people to believe they're the same species at different ages.

To estimate the life stage of an animal you're observing, you need to go through all the traits in this chapter and understand that the markers of a newborn may not be the same as those of a juvenile, and both may vary significantly from the adult. Let's use the hairy woodpecker as an example. Woodpeckers are altricial species, meaning that they are born naked, blind, and completely dependent on their parents; this contrasts with precocial bird species like ducks, which hatch out of the egg with a full coat of feathers, open eyes, and the ability to walk and swim. So, a newborn hairy woodpecker must remain in the nest for the first four weeks of its life, and it will have an incomplete set of feathers for most of that time—something that certainly doesn't describe the adult downy woodpecker mistaken for a young hairy.

Once it leaves the nest, it is considered a fledgling, and because birds tend to grow very quickly, it is almost as big as its parents; it's also already larger than the adult downy woodpecker. However, at this stage of its life, its feathers are still fairly scraggly, especially around its head. Once the last of the baby feathers have been molted, the overall pattern resembles adult plumage, but with an important difference: Instead of bright-white feathers to contrast with the black markings, theirs are light brown. Male hairy woodpecker fledglings have also not developed the bright red patch on the back of the head that will mark their maturity, a sex characteristic also shared with downy woodpeckers.

So, a few weeks after this, we have a male juvenile hairy woodpecker, who is now as big as mom and dad and has the same pattern of black feathers. Yet those brown feathers mark him as being just a teenager, and should he happen to try to woo a female, she will instantly know that he still has some growing up to do. Only when he grows old enough to shed the brown feathers, grow in white ones, and display the red male badge on his head will he stand a chance with a prospective mate.

Finally, you're never going to see an adult hairy woodpecker feeding an adult downy woodpecker. Woodpeckers do feed their young for a few weeks after they leave the nest, but in that period the juveniles are still obviously different from the adults. By the time the juvenile grows its adult plumage, it's long past the age where it's still chasing after its parents begging for one more meal before they cut the proverbial apron strings.

Unfortunately, other than "younger members of a species are generally smaller than adults," there really aren't any universal markers of age across all animals. What works for estimating the life stage of primates won't necessarily hold true for cephalopods. If you can at least narrow what you've found down to a certain taxon such as class or order, you may be able to do some research on typical developmental markers of species in that group, which might offer some clues. Otherwise, just continue to observe and record as much as you can about the animal in the hope that you'll be able to use the information later to determine its approximate age.

Sex

This is another one of those traits that won't always be immediately obvious. Yes, there are some sexually dimorphic species in which it's clear which is which; the male Indian peafowl (*Pavo cristatus*) is the classic peacock with vivid blue-and-green coloration, while the female is a much more subdued brown and white. However, in some dimorphic species, the differences between females and males may be subtle; many female snakes are larger than males, but if you don't have a male for comparison and if the size range of the sexes overlaps, you may not realize you have a female unless she's exceptionally long. Many animals are not sexually dimorphic at all, meaning that every adult member of the species looks the same regardless of sex. And, of course, there are species that are hermaphroditic, in which every individual has both female and male sex organs.

Which means it is entirely okay to record the observed sex as "unknown" when you just aren't sure. That includes dimorphic species if you aren't familiar with which set of traits goes with the female versus the male. There are cases where scientists thought they were looking at two sexes of a newly discovered or seldom-observed organism, when in fact they were actually two separate *species* (and vice versa, for that matter).

What happens if you have a group of varied animals and you aren't sure how many species are actually present? Watch how they engage with one another, if at all. In social species and mated pairs, the interaction between two individuals with distinct appearances may be a clue that they are of the same species. As with adults and juveniles, you won't generally see members of different species interacting in a social, friendly way together. While multiple species may show up at the same watering hole or food source, the only interaction is likely to involve territoriality over the resource, not cooperation or other positive social behaviors, and the rest of the time they ignore one another.

Movement

Unless an animal is resting, sleeping, sick, or dead, you're likely to see some movement. Start with the basic type of locomotion—such as walking, running, swimming, flying, or crawling—and then be more detailed in your description:

- The bird flew in undulating patterns, alternating between flapping its wings and gliding.
- The snake I startled curled up in what looked like a defensive pose; when I moved away, it turned around and quickly slithered into the grass.
- I saw a black beetle trundling along the ground at a steady pace, as though it had someplace important to be, and no rocks or dead leaves were going to get in its way.
- A harbor seal steadily swam a little ways off the shoreline, ducking its head under periodically, only to resurface several feet away and resume its journey downshore.

undulating "flap-flap-glide" flight pattern - wings clamped tightly to sides during glide

"no 'mustache' = female"

"Red shafted" western variant w/ red-orange underside wings/tail

You might draw what the animal looks like while moving as opposed to at rest, or try to depict the way it moves as a series of lines or shapes. Simple line drawings are fine; you don't have to be super elaborate if you don't want to. You're just taking notes to help you recall the movement later.

One thing you should never ever do is touch, scare, or harass an animal at rest to make it move. Not only does this stress the animal, but it could potentially cause it to hurt itself or you if it panics and tries to escape. Keep in mind that you are trying to observe natural behavior, not a forced interaction. Wild animals do not exist for our entertainment and should be respected in their homes.

Sounds

Unlike plants and fungi, the animal kingdom emits a full chorus of barks, howls, songs, screeches, clicks, hisses, and many other varied sounds. Many of these are within our hearing range, although some, like the high-frequency squeaks of bats and low-register sounds made by whales and elephants, can be detected only with special equipment.

It's possible during your adventures in nature that you'll hear animal sounds but not be able to see who made them. Many birds and other small vertebrates have alarm calls they use to warn others when danger is near, which, in their opinion, includes the everyday naturalist. But if you're quiet—and lucky—you might also get to hear mating calls, encouragement to young, and other social chatter. Don't be surprised if whatever you're listening to quiets or stops entirely when you approach; while sounds have important functions in nature, they can also be a liability as they reveal an animal's location to predators looking for a meal.

Some animals have had to adapt their vocalizations to the ever-encroaching presence of humans. Birds may have to sing at a higher volume to be heard over our noise. Whales also have had to increase their volume, but because of overwhelming noise pollution, they simply have fewer opportunities to speak to one another. The family of coyotes (*Canis latrans*) that den on the farm I live on sing together as a form of social bonding, but—like domestic dogs (*Canis familiaris*)—they now also howl in response to ambulance and fire sirens on the road.

Not all animal sounds are made with mouths. Crickets (Grylloidea spp.) "sing" by rubbing their wings together. Some species of hummingbirds (Trochilidae spp.) perform dive-bombing displays, and a few of them can produce a high-pitched chirp with their rapidly vibrating tail feathers. And click beetles (Elateridae spp.) get their name from the sound they make with special structures on their exoskeleton.

Some species have quite a varied repertoire! Many wild canids, for example, have a whole vocabulary of howls, growls, whines, and snorts they use to communicate with one another. Different populations of a species may have regional dialects, unique enough that they might be mistaken for completely different species. Certain birds will not only make the calls and other sounds commonly associated with their own kind, but also will mimic other noises they hear in their environment; the superb lyrebird (*Menura novaehollandiae*) is a particularly notable example that makes spot-on imitations of chainsaws, motors, and other machinery. If you see an animal you recognize making a sound you've never heard before, you probably just haven't had a chance to listen to everything it has to say, so look for some recordings online and see if any of them match.

When possible, make a recording of whatever animal sounds you hear. Most smartphones these days have decent microphones on them, and you can use either the built-in camera to take video or a more dedicated audio app. This will suffice for most people, but there's nothing wrong with having a more professional and robust portable recording system if you like.

Playing recordings of certain animal calls to get the live ones to respond is another one of those highly contentious topics among naturalists. If you are a professional researcher or a citizen scientist helping with a project, playbacks may be justified. For the most part, though, the rest of us don't need to get birds or other animals stressed out looking for a potential interloper in their territory just so we can catch a glimpse of them.

If you're caught without recording equipment while hearing something interesting outside, do your best to listen closely as long as the animal keeps calling, and then record your impressions as soon as you can afterward. You might try to write out what it sounded like; some birds have common mnemonic phrases, like the barred owl's (*Strix varia*) "Who cooks for you? Who cooks for you-all?" or the "teakettle, teakettle, teakettle" of the Carolina wren (*Thryothorus ludovicianus*). But you might just have to make up something that sounds right, such as the "huff" of an upset black bear (*Ursus americanus*) or the "zeeeeeee-ooooooooooo-zeeeeeeee-ooooooooooo" of cicadas (*Neotibicen* spp.) on a summer night.

If you're not particularly self-conscious, you can try recording yourself attempting to imitate the sound you heard to the best of your ability; you can sometimes get surprisingly recognizable results! I can't easily match the range or sharpness of the call of a northern flicker (*Colaptes auratus*), but I can make a "wuk-wuk-wuk-wuk" sound with an even rhythm and a rising and falling pitch. I can also demonstrate how it differs from the similar but uneven staccato "wuk—wukwuk-wuk-wuk-wukwuk—wuk" and lower pitch of the pileated woodpecker (*Dryocopus pileatus*).

Usually, the resemblance between two close-but-not-quite sounds is unintentional. On one of my visits to the Konza Prairie in northeastern Kansas, I heard what sounded like a rattlesnake's buzzing tail about ten feet off the trail. I was pretty sure it was one of the several dozen grasshopper species that make their home there, but I did contact prairie managers at Kansas State University to see if there was any chance of an errant rattler there. (There was not.)

However, nature does have its deliberate mimics. Steller's jays and blue jays both will imitate red-tailed hawks with uncanny accuracy, fooling even experienced birders. They usually do this to let other birds know a hawk is near, but they'll also use the hawk call to scare other birds away from food they want!

Other Behaviors

Vocalizations might be the most obvious behaviors you'll notice even before you see an animal, but if you get to watch one for more than a few seconds, it can be a great opportunity to see them seeking food, finding mates, caring for young, preening, making and maintaining a home, and so forth. Some behaviors can be quite distinctive and help with identification; hog-nosed snakes (*Heterodon* spp.) have a rather impressive death-feigning display to warn off predators, while archerfish (Toxotidae spp.) hunt insects near the water's edge by spitting streams of water that knock them down into their reach.

You might not always know exactly what's going on, but do your best to watch what the animal is doing, where it goes, how it's interacting with other living beings and its habitat, and so on. Video is always useful when possible, but keeping detailed notes may also make it easier for you to research the behavior later. Be as descriptive as you can, both with regard to the behavior and its context; instead of writing "I saw a brown and white bird walking by the water," you might instead note "3:10 p.m.—saw a lone brown and white shorebird with brown spots on its breast, about jay-sized or a little larger, appeared to be foraging along the Deschutes River by probing the gravel with its beak. Bobbed its tail up and down as it walked in a stop-and-go pattern. After about ten minutes, it was startled by people coming around the bend and flew upriver with rapid wingbeats."

Interactions among more than one animal are often especially exciting. If they're all the same species, you may get to observe interesting social behaviors, like grooming each other, fighting, caring for young, or mating rituals (other species generally aren't as private about that as we are!). Interspecies activity often involves some sort of conflict between predator and prey, although there are also cases where two predatory-scavenger species may work together to acquire food. Even squabbles between multiple bird species at a feeder can be quite dramatic!

Do your best to not anthropomorphize animal behavior; that means don't assign human motivations or biases to other species. Keep your observations as neutral as you can, although it's okay to record your own reactions to them. For example, "I marveled as a young snowshoe hare (*Lepus americanus*) briefly sniffed a newborn mule deer fawn (*Odocoileus hemionus*) hiding in the grass where its mother had left it" is a more accurate interpretation than "I saw a baby bunny and a baby deer sniff each other and that must mean they're friends, just like Bambi and Thumper!" The latter is not too far off from similar comments I've seen online about photos of two wild species in proximity that were predator and prey who were definitely *not* going to be the best of buddies.

In many cases, you won't have much of a choice in how long you can observe the animal. They're often the ones who decide to head off out of sight, even if you were being as quiet and calm as possible. But don't feel pressured to stand for hours watching a rabbit chowing down on clover. Wildlife viewing may be a privilege we get to enjoy, but we aren't obligated to be full-time witnesses.

Do, of course, pay attention to your safety as well as that of the animals. There are certain things you just don't want to get too close to, like a wasps' nest or a herd of bison. Don't be so focused on watching that cedar waxwing (*Bombycilla cedrorum*) that you fall into the river nearby. And never force an interaction with an animal to "make it do something." Any time an animal is not moving around, it's still engaged in a behavior; the hawk might be scanning the ground for prey, or the vole may be hiding from the hawk (and you).

Tracks, Scat, and More

Sometimes all you get of an animal is a sign that it was there earlier—a line of tracks, a shed feather or a tuft of fur, a pile of scat. Once you learn to recognize the signs of a particular species, they add to your understanding of its natural history.

Tracks

Muddy areas are the best place to start looking for animal tracks year-round, and if you get snow in the winter, it's likely temporary tracks will be there as well. Dry sand and gravel don't record prints as clearly, nor will ground covered in heavy vegetation. Tracks won't always be completely clear, such as when the animal is sliding in sloppy mud or stepping in its front footsteps with its hind feet as it moves. All tracks will degrade over time as wind, water, and other animals erode or cover them, making them harder to discern.

Tracks can be quite tricky to identify, so I recommend getting a tracking book or two relevant to your area. Some are quite general, focusing on the most common mammals of North America, for example. Others may be more specific to a region or even branch out into birds, reptiles, and amphibians. There are also online groups on Facebook and elsewhere dedicated to animal track identification.

Pay attention to the size and shape of the tracks. Determine whether you're seeing impressions from one or multiple feet; for example, a rabbit leaves a set of four round divots in a rough Y shape that people often mistake for one single footprint. If all the prints you see are the same size, especially if there's just one line, it's likely you're dealing with one animal. Tracks of multiple sizes and shapes indicate several individuals, possibly of more than one species. If they have significantly eroded and have softer, less distinct edges, they may be larger than when they were first made, so the animal may have a smaller foot than the marks may seem to indicate at first.

Top, left to right: cougar; crocodile | Middle, left to right: elephant; moose |
Bottom, left to right: raccoon; wolf

When assessing a particular track, try to imagine the paw, foot, tail, or wing that would leave that mark. Mammals tend to have paws with pads on the center and each toe (or hooves, in the case of ungulates). Birds usually have three long, skinny toes that splay out at wide angles as they walk; some have a fourth toe that extends backward. If a bird takes off, it may leave distinctive wing prints behind. The prints of some amphibians and reptiles also have marks from their tails or bellies dragging the ground, and snakes, of course, have no feet at all. The tracks of insects, spiders, and other small invertebrates often look like a double line of tiny divots, again sometimes with a belly drag mark between.

Now and then a photo will circulate online showing a series of deep, round prints in the snow, and then they stop right at a set of wing prints. These pictures are often erroneously labeled as evidence of a raptor of some sort catching a rabbit, grouse, or other prey animal on the wing. However, in that case, the raptor's impact with the prey would leave a messier cavity in the snow and might leave feathers, fur, or blood behind. In fact, the wing prints are from a ptarmigan (*Lagopus muta*) walking through the snow before taking flight.

A single paw print that is four inches across and five inches long would have to be from a sizeable animal, so you can definitely rule out a rabbit! If you can see individual toes, count how many there are. You might not know how many toes every mammal in your area has, but if you know that felines and canines generally have four toes per foot, they're not going to be the ones leaving a five-toed print unless they have a rare case of polydactyly. The two-toed hoofprints of a white-tailed deer (*Odocoileus virginianus*) and an elk (*Cervus canadensis*) may have a similar shape, but the elk's will be larger and heavier; even the track of an elk calf is often more robust than that of a full-grown white-tailed buck.

The distance between the tracks indicates the animal's stride length. You may not be able to determine how fast it was going, especially if you aren't familiar with each species' gait pattern—in what order and pattern its feet land when walking, trotting, running, and so on. But you can at least estimate the size of an animal that could take a step that long. Even when slowly grazing, a deer is going to take a stride that is much longer than that of a raccoon (*Procyon lotor*).

Take a few pictures of the prints you've found. If the tracks are in a place that is relatively undisturbed and you have a couple of hours to spare, there are directions online on how to make a plaster cast of it, so that you have a permanent three-dimensional record.

Fur, Feathers, and Other Bits and Parts

Some animals leave behind physical remnants of themselves. These are usually small, like bits of fur, molted feathers, or a reptile's shed skin. Occasionally you may run across an antler from a species of deer or bones from an animal that died some time ago.

Again, consider the size and scale of the animal that may have left the remnant behind. The individual hairs in a muskrat's coat are going to be shorter than those of a coyote, and the shade of brown is likely to be different as well. A fox sparrow (*Passerella iliaca*) will have smaller feathers than an American bittern (*Botaurus lentiginosus*), and both the shape and color patterns of their feathers are quite different. The ten-inch-long red-bellied snake (*Storeria occipitomaculata*) won't shed a three-foot-long skin.

When you find bones, remember that the live animal was also covered in flesh, skin, hair or feathers, and so on. The bones are just at the center of all that mass. Say you find an animal skull; imagine how much bigger it might be with all that muscle, sinew, and skin on it, and then what animal might have a head about that size and shape. Try holding the skull at the approximate distance off the ground you think the animal's head might have been when it was walking; this might give you some ideas of how large the animal was.

Skulls tell us a lot about an animal, particularly its diet. Carnivorous mammals tend to have sharp teeth, herbivores often have flat molars (the teeth in the back of the mouth), and omnivores are somewhere in between. All the teeth in a lizard's mouth look the same, and a bird, of course, has no teeth at all. Other bones tell stories, too; the long hind leg bones and shorter front limbs of frogs and toads point to their hopping and jumping locomotion. Hollow bones are generally those of birds; not only are they lighter for flight, but they are integral to how the bird breathes.

Arthropod exoskeletons can persist long after death, as can the shells of mollusks. It's quite common to find an entire, undamaged dead insect; instead of trying to guess what you have from a single leg or wing, you have the whole animal to work with.

Some of the remains you find may be left over from a predator's kill or an animal that died of other causes. Think about what animal might have hunted the one you found. How large a predator might it take to kill it, and what smaller scavengers may have stopped by afterward? Sometimes where you find the remains can be an important clue. Raptors often take their prey into a tree, and you'll often find a pile of feathers or fur on the ground beneath the branch where they dined. Some animals, like coyotes or wolverines (*Gulo gulo*), may cache a kill by burying it under soil or debris. If you find a bigger carcass like that of a deer up in a tree, it's likely that a large cat, such as a cougar, dragged it up there.

Herbivores often leave remnants of their meals lying around, too. If you find a pile of scales from conifer cones at the base of a tree, look up and you might see a squirrel extracting the seeds from a cone. Look closely at where deer have been, and you may see where they cropped grass or other foliage. Beavers (*Castor canadensis*) leave characteristic stumps with pointed ends when they cut down trees.

Scat

This is just another term for an animal's fecal matter; *frass* is a term specifically for the feces of insects. The size, shape, and contents of scat all make good clues as to who left it there. There is going to be some variation in scat size, even from an individual animal, so think in terms of averages. If you have a dog, cat, or other pet that you clean up after, think of their scat in relation to their size as good examples. There are, of course, outliers; frogs, for example, can produce a pile of scat that is much larger than you would expect from such a small animal!

The shape of an animal's scat may vary depending on what it's been eating and how healthy it is. Elk scat can be quite wet and "ploppy" in spring when the herd is eating lots of fresh green grass, but in winter it tends to be piles of round pellets because of the drier forage at that time of year. A coyote's scat is normally cylindrical, similar to a dog's, but if the coyote has some sort of gastrointestinal disease or parasite, it may end up with looser stool or diarrhea.

Not all scat will be like a mammal's brown, relatively firm droppings. Birds have only one waste orifice, the cloaca, and their scat is a mix of both feces and urine, which gives it its unique dark-and-light coloration. You may find other signs of their diet, such as undigested seeds or red or purple splotches from berry juice.

Some predatory birds, like owls, vomit up pellets containing bones, hair, and other indigestible parts of their prey. While this isn't true scat, it can be used in the same way to help with identification. The pellet regurgitated by a great blue heron (*Ardea herodias*) will have more aquatic prey remnants like crayfish shells or fish bones than the pellet of a barn owl (*Tyto alba*). Some people like to dissect bird pellets to examine the bones and other parts within, identifying the prey animals as well as the predator.

Scat changes over time, drying out after just a few days in summer and sometimes getting mushier and softer in rainier times. Detritivores like dung beetles (Scarabaeinae spp.) may start feeding on the scat shortly after it has been dropped, and bacteria and other decomposers also work on breaking it down. As it decays, it may be more difficult to identify, so the fresher you can find it, the better.

Because naturalists are the sort of people who want to know everything we can about nature, there are thankfully several field guides and at least one Facebook group dedicated to identifying animal scat. Any skilled animal tracker will also have learned to identify scat as well as tracks, so taking a class in tracking might help with your scat IDs as well.

Scratches and Other Marks

I mentioned earlier that beavers might leave signs of feeding on nearby trees. But there are other marks you may find, too. When new antlers first grow on deer, they are living bone covered in a thin, fuzzy skin called velvet. Once the antlers are fully grown, the bone dies, and the velvet dries up and sloughs off. To facilitate the velvet's removal, the deer scrape their antlers against trees, rocks, and other hard surfaces, often leaving visible scars behind.

Other scratches on trees may be from bears or cougars marking their territory, climbing and slipping, or just getting a nice stretch in. In this case, you will generally see distinct lines from the individual claws, although they may overlap. Large areas of missing bark without claw marks could be where large animals have been using the tree trunk as a convenient back scratcher.

When the BBC was working on its television show *Planet Earth II* a few years back, it published the most delightful video on its YouTube channel: "Bears Dancing to 'Jungle Boogie.'" It features up-close-and-personal footage of wild Canadian grizzly bears getting in some satisfying back scratches on trees while the funky sounds of Kool & the Gang provide the perfect soundtrack.

Habitat

A habitat is simply the place where a given animal (or plant or fungus) lives, including any nest or den and its overall territory. An ecosystem is made of up all the habitats of all the living beings within its bounds. Habitats overlap; the habitat of one plains zebra (*Equus quagga*) living in northern Botswana may have the habitats of multiple meerkat (*Suricata suricatta*) families within it.

You won't necessarily know the boundaries of a given animal's habitat, but you can find signs they were there. In addition to tracks, scat, and the like, look for places where the animal may sleep or raise young. Some have been built by the animal, like a bird's nest, or taken over after another species abandoned it, such as a burrow that may see several different types of animals take residence over the years. Animals often take advantage of natural shelters, too; caves, thickets, and spaces under rocks are just a few places where they might find a convenient home.

A Few Final Notes

As I've mentioned before, observing animals in the wild can be more difficult than sitting down with a plant or fungus for a good, close look. Their tendency to move suddenly out of sight means we may not get more than a few seconds to watch them and gather as much information about them as we can. While many of us may have opportunities to visit some wildlife species in a zoo or aquarium, captivity often changes animal behavior, and, in some cases, other physical traits. Plus, if your goal is nature identification, captive animals will already be identified and labeled for you, so they're less of a challenge.

Wildlife photography is certainly an option if you're interested in pursuing it. All the photographers I know say that you have to have tremendous patience and be willing to spend a lot of time outdoors looking for photo opportunities. Even if you see an animal, it doesn't mean you'll be able to get a good shot of it (although you might get some great bloopers to share in the Crap Wildlife Photography group on Facebook!). But the effort is often worth it when you get that clear picture displaying an animal at its best.

There is a code of ethics to follow when observing wild animals, whether you're photographing them or not. First, keep your distance as much as possible; binoculars and spotting scopes—and telephoto lenses for cameras—are your friends. If the animal is resting or otherwise not moving around, don't make sounds or other disturbances to agitate it.

Do not use food, recorded calls, scents, or other methods to attract wildlife, as it can make them dependent on humans and change their behavior for the worse in several ways. Songbirds who show up at feeders are the exception because they are less likely to become habituated as long as you aren't trying to hand-feed them, although some people feel that even bird feeders are unethical. And, certainly, do not trap or otherwise contain an animal just so you can observe it unless you are part of a formal scientific study with professional wildlife handling procedures designed to minimize stress and harm.

If you observe a rare species, it may be best to put any photos you have on iNaturalist with an obscured location rather than exact GPS coordinates and contact your local fish and wildlife department to let them know what you saw. Too often, people crowd to a place to see a rarity, which can cause it distress or even lead to it leaving the area.

Finally, if you are concerned that an animal may be injured or ill, contact the closest wildlife rehabilitation center; or again, your local fish and wildlife department may have suggestions on who can offer aid. Unless you are a licensed wildlife rehabber or similar professional or acting under their direction, do not remove an animal from the wild, even to save its life. Death is a natural part of life, and while it can be heart-wrenching to watch some-thing suffer, its death means that scavenging animals and other organisms will get much-needed food to live another day, and the deceased animal will return to the never-ending cycle of nutrients in the ecosystem in which it was born and died.

Done ethically and safely, wild animal observation and identification can be incredibly rewarding, and for many people, the extra effort makes success even sweeter. Should you not have the patience to try to identify a moving target, there are more stationary beings to focus on. In the next chapter, you'll learn more about identifying something that won't run away from you—plants!

Templates and Case Studies

I've created a few templates to help you keep track of the traits just discussed. Blank versions of these templates that you can copy and use either out in the field or once you get back home are on pages 246–251. I've also included sample templates here and in the next two chapters to show you how you might use them to record your observations in detail.

After filling in the template, I usually write up a more detailed case study as a record of a specific species I've observed in the wild. (My initial notes are not usually this polished as they are here; I've organized them for the purposes of this book.) While you are certainly welcome to use the following as a guide for your own field notes, it's okay if you just want to use the templates to jot down some basic information for each new species you encounter.

ANIMAL IDENTIFICATION TEMPLATE

Species Observed:
Virginian tiger moth (*Spilosoma virginica*)

What, When, and Where?

DATE: September 4, 2022

LOCATION: Home, Long Beach, WA, in driveway near west barn

NUMBER OBSERVED: 1

LIFE STAGE: Larva (caterpillar)

SEX: Unknown

ALIVE OR DEAD? Alive

HABITAT (be as detailed as possible):
Observed on gravel driveway, but species is typically found in grassy areas and among other low-profile vegetation.

Observed Behaviors

MOVEMENT: Crawling quickly with undulating movement

SOUNDS: None observed

OTHER: None observed

What's It Look Like?

VERTEBRATE OR INVERTEBRATE?
Invertebrate

SIZE/PROPORTIONS:
Average size, about 1½ inches long, maybe ½ inch in diameter at the widest point. A little more slender than some "woolly bears" I've seen.

OVERALL SHAPE/DETAIL SHAPES:
Typical fuzzy caterpillar with hair all over, a long cylinder with six tiny legs and a small, round black head almost entirely obscured by black hairs.

COLORS/PATTERNS: Mostly rust or cinnamon color, but the front ¼ inch was black. No black on the rear end.

TYPE OF OUTER COVERING (feathers, scales, exoskeleton, etc.):
Covered in many very fine hairs, with a fuzzy "woolly bear" appearance. Some hairs significantly longer than others.

VISUAL TEXTURE:
Again, "fuzzy" about covers it.

TRACKS, SCAT, NESTS, ETC.:
None observed.

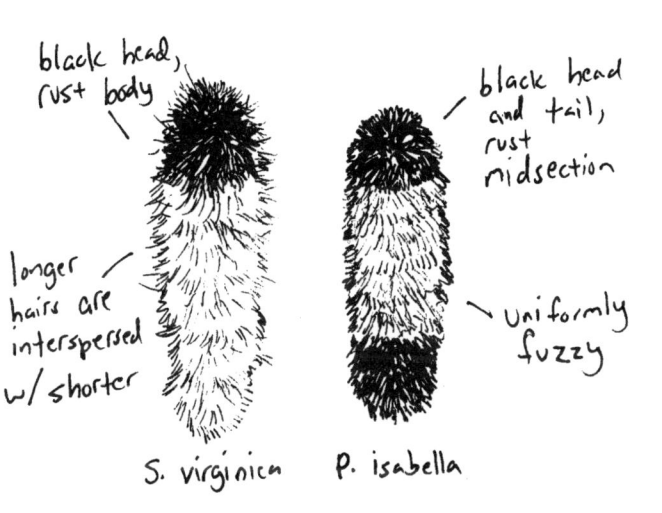

black head,
rust body

longer
hairs are
interspersed
w/ shorter

black head
and tail,
rust
midsection

uniformly
fuzzy

S. virginica   P. isabella

CASE STUDY

Species: Virginian tiger moth (*Spilosoma virginica*)

DATE: September 24, 2022

LOCATION: Home, Long Beach, WA, in driveway near west barn

OBSERVATION NOTES: Saw an unusual caterpillar this morning. While typical coloration for an Isabella tiger moth (*Pyrrharctia isabella*) consists of a black caterpillar with a rust-colored band of varying width around the middle, this one was almost entirely rust with black only on the front quarter inch or so. It was about an inch and a half long, not as thick as some that I've seen. It was hurrying along, full speed ahead, with the typical undulating crawl of this species. I've seen several in recent weeks, and it's the season for them to be eating as much as they can before going dormant later in fall until next spring. While this one was out on the gravel driveway, I also see them frequently in nearby pastures and meadows.

DISCUSSION: "Woolly bears" are among the most easily recognizable caterpillars in North America, and many people believe that the width of the rust band around the middle of an Isabella tiger moth caterpillar foretells what the winter weather to come will be like. But what happens when you run across one that looks unusual? I was pretty sure this fuzzy little critter was an Isabella tiger moth caterpillar, but I wanted to be sure. I'm familiar with several other species of tiger and tussock moths in the area, all of which had similarly hairy larvae. However, none of them typically have the black-and-red coloration of the Isabella tiger moth.

I already ruled out the spotted tussock moth caterpillar (*Lophocampa maculata*) because, although they can have the same black-and-red coloration, they also have prominent tufts of white hairs on their heads and hind ends. The rusty tussock moth (*Orgyia antiqua*) is black with thin rust bands and yellow spots on its back half, while the silver-spotted tiger moth (*Lophocampa argentata*) is yellow with black stripes down the sides and long chestnut hairs.

However, two separate people on iNaturalist insisted that what I had seen was the Virginian tiger moth. Their caterpillars may be found in a wide array of shades. I've seen white, yellow, and rust versions, although none of them had any black on them. In looking through other people's photos on iNaturalist, I did occasionally see a rust caterpillar with a black head identified as the Virginian tiger moth, so it was worth investigating further.

Upon closer examination, I noticed that the Virginian species' caterpillars have a scattering of longer black hairs all over their bodies, even on the reddish portions. The Isabellas, on the other hand, have more or less uniform hairs all over. So, I took another really good look at the photo of my specimen. Sure enough, there were the longer hairs, marking this as my first black-headed rust variation of the Virginian tiger moth caterpillar.

ANIMAL IDENTIFICATION TEMPLATE

Species Observed:
Osprey (*Pandion haliaetus*)

What, When, and Where?

DATE: April 15, 2023

LOCATION: Captain William Clark Regional Park, Washougal, WA

NUMBER OBSERVED: 2

LIFE STAGE: Adults

SEX: 1 female, 1 male

ALIVE OR DEAD? Alive

HABITAT (be as detailed as possible): Sighted in active nest built on wooden platform on tall pole. Situated near a line of trees in developed park in urban area just north of the Columbia River.

Observed Behaviors

MOVEMENT: In flight, broad, powerful wingbeats quickly achieving speed

SOUNDS: Chittering calls

OTHER: Active courting and mating behavior in nest

What's It Look Like?

VERTEBRATE OR INVERTEBRATE? Vertebrate

SIZE/PROPORTIONS: Large hawks, about 30 inches long from end of beak to end of tail, wingspan over 5 feet across. Female larger than male. Typical hawk proportions, robust but not heavy.

OVERALL SHAPE/DETAIL SHAPES: Typical large hawk with streamlined body; powerful large wings; broad fan-shaped tail; sharp, curved talons; short, sharp beak; and small, rounded head with large eyes.

COLORS/PATTERNS: Typical adult plumage with white body and head, black beak, dark brown band across yellow eyes, and upper side of wings dark brown; wings and tail have white-and-brown-striped flight feathers. White feet, black talons.

TYPE OF OUTER COVERING (feathers, scales, exoskeleton, etc.): Feathers all over body, feathers cover entire leg with only feet bare.

♀ larger then ♂

Columbia River ~200 ft. south ←

Platform app. 30' tall

VISUAL TEXTURE:
Sleek, smooth feathers ruffled only by the wind. Looks like they would be soft to touch (if I wanted to lose a few fingers)!

TRACKS, SCAT, NESTS, ETC.:
Nest made of various sticks and other woody debris, large (about 7 to 8 feet across), and built on large wooden platform on tall wooden post made specifically for raptor nests.

CASE STUDY

Species: Osprey (*Pandion haliaetus*)

DATE: April 15, 2023

LOCATION: Captain William Clark Regional Park, Washougal, WA

OBSERVATION NOTES: Stopped here on a road trip and just happened to see a pair of ospreys actively mating in a nest on an artificial stand at the park. The male would stand on the edge of the nest while the female was settled in the center. Every so often, he would then hop onto her back, and they would mate, chittering at each other the whole time. This went on for a good fifteen minutes before the male flew off toward the river, followed shortly by the female; both had powerful wingbeats. The female was a little larger than the male. Both had typical osprey coloration, a white head, and a brown stripe across the yellow eyes with a black beak. White body, dark-brown upper side of wings, and wings and tail show white-and-brown-striped flight feathers. In flight, the wingspan has a distinctive M shape as viewed from underneath as the wings are bent in the middle at the wrist. Large, sharp curved talons and short curved beak; typical hawk shape and proportions.

DISCUSSION: There's really nothing quite like an osprey. No other raptor in North America has the same sort of brown-and-white coloration with the distinctive bands over the eyes, and few of them are as large. Adult bald eagles (*Haliaeetus leucocephalus*) are brown with a white head and tail, but they lack the osprey's white belly and facial markings and do not have stripes on the flight feathers. They are also larger than ospreys, although this can be difficult to discern if you don't have both species near each other for comparison. Some light-phase red-tailed hawks may look superficially like an osprey with a white belly and striped flight feathers, but they generally have a reddish-brown head with a white throat, and adults have the characteristic red tail.

The most common confusion comes from seeing ospreys in flight, especially at a distance. In my area, they may be seen flying overhead at the same time as bald eagles and turkey vultures during breeding season. However, neither eagles, vultures, nor red-tailed hawks have the M-shaped wingspan of the ospreys. Moreover, turkey vultures hold their wings in a V shape, rather than flat and level, and they tend to teeter-totter side to side as they soar, while ospreys do not. Finally, when ospreys catch a fish, they carry it head- or tail-first, while eagles carry their fish with the side facing forward.

CHAPTER 6

How to Identify Plants

bromeliad (*Tillandsia tequendamae*)

WHEN I TEACH IN-PERSON NATURE IDENTIFICATION classes, plants are always the easiest subjects to study. They stay in one place for convenient observation, many of them have distinctive features that help with identification, and while many plants die back in the winter, most places have some perennial species to explore year-round.

The challenge of plant identification is that there's a lot of specialized botanical vocabulary to learn, especially if you want to get more specific than "a plant with green leaves and a straight stem." I'll cover a selection of terms here, although this chapter should not be seen as a complete introduction to botany. If you don't recognize a particular word, either consult the glossary on page 252, search for it online, or look it up in *Plant Identification Terminology: An Illustrated Glossary* (see Recommended Resources on page 257).

Certain groups of plants are notoriously difficult to identify down to the species level, even for experienced botanists using dichotomous keys. The grasses are a good example. Most grasses have long, slender leaves (colloquially known as blades), and those of related species may look extremely similar. The flowers and seeds are more distinctive, but they're available for study only part of the year, and even then, you can have look-alike species that are difficult to distinguish.

Like the animal kingdom, plants are also quite varied in form. Among the few features common to all plants is the cellulose that makes up the walls of their cells and the chloroplasts needed to photosynthesize—the process of turning sunlight into food with the use of chlorophyll. Yet that is where the universal commonality ends. A massive coast redwood tree (*Sequoia sempervirens*) towering three hundred feet spreads its roots beneath the rhizomes of a carpet of moss a fraction of an inch thick. Instead of drawing nutrients and water from the soil, *Tillandsia* air plants use their roots to cling to perches high up in a forest canopy and absorb what they need as it flows by on the wind. Parasitic species such as ghost pipe (*Monotropa uniflora*) siphon their food and drink from their neighbors and have lost their chlorophyll due to disuse, although they are still undeniably plants.

Still, you can use some common features of plants to identify most species; we'll explore them after a quick look at how plants are put together.

Basic Plant Anatomy

The first thing you need to consider when you are looking at a plant is whether it is vascular or nonvascular. The vascular system is a series of tissues—xylem and phloem—that move water and nutrients through the plant's body. Because it is highly efficient, it allows vascular plants to grow taller and larger than nonvascular plants, as well as have more complex structures like true roots and leaves. Nonvascular plants are limited in how well they can move resources around, and so they are generally smaller and simpler. Rather than roots, they have rhizoids, as well as leaf-like structures called phyllids that are only one cell thick and lack some of the structures found in a vascular plant's leaves.

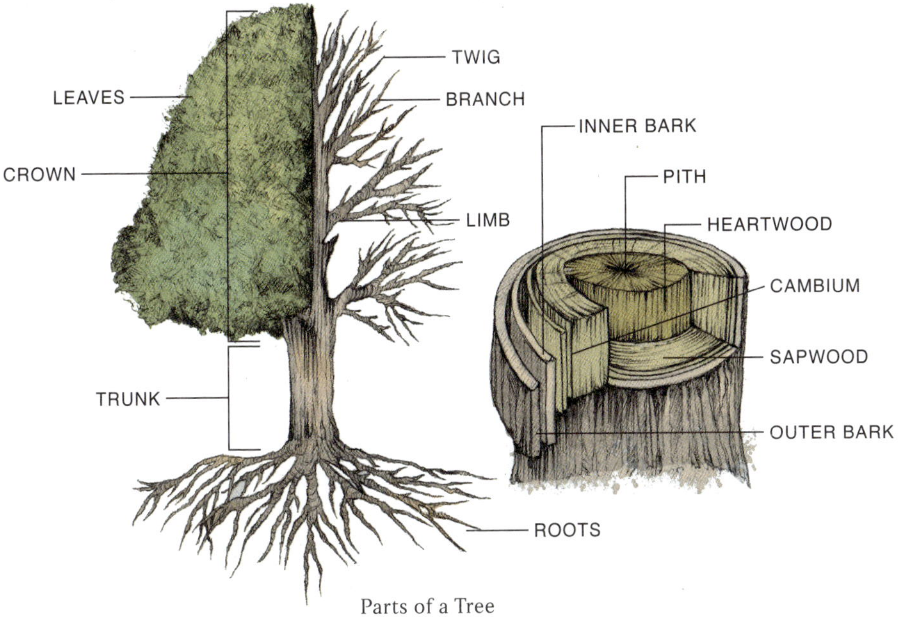

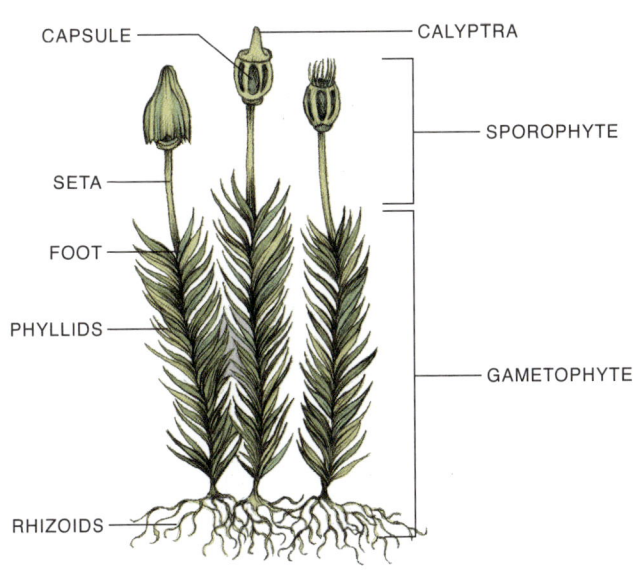

LEAVES

TWIG

BRANCH

CROWN

INNER BARK

PITH

HEARTWOOD

LIMB

CAMBIUM

SAPWOOD

TRUNK

OUTER BARK

ROOTS

Parts of a Tree

CAPSULE

CALYPTRA

SPOROPHYTE

SETA

FOOT

PHYLLIDS

GAMETOPHYTE

RHIZOIDS

Moss

Most terrestrial plants are vascular. The bulk of nonvascular plants are known as bryophytes and include mosses (Bryophyta), liverworts (Marchantiophyta), and hornworts (Anthocerotophyta), with green algae making up the rest. They can be differentiated from vascular plants by how their vascular tissue is arranged (something you won't see without a microscope) and how they sexually reproduce. Plants have two stages to their reproductive cycle; one stage produces a haploid sporophyte with a single copy of the parent's chromosomes, while the diploid gametophyte created in the second stage has two copies. Whereas most vascular plant sexual reproduction is primarily via gametophytes (seeds), all the main reproductive stages of nonvascular plants involve sporophytes (spores).

A few groups of vascular plants produce spores, such as ferns (Polypodiopsida), clubmosses (Lycopodiopsida), and horsetails (*Equisetum*), but they are exceptions to the rule. Some plants, both vascular and nonvascular, may additionally be able to reproduce asexually through fragmentation, gemmae, cloning, and so on.

Green algae is a group containing multiple clades of nonvascular plants. The Chlorophyta are the best known, but other taxa such as Charophyta and Streptophyta are also included. Some scientists argue whether certain groups categorized as green algae should be considered protists instead; *protist* is shorthand for anything that isn't an animal, plant, or fungus. Red algae (Rhodophyta) and brown algae (Phaeophyceae) also find themselves shuffled between the plants and protists, so if you ever hear someone say that "algae aren't really plants," they may or may not be correct, depending on what current evidence suggests and who's making the argument.

The bulk of vascular plants produce seeds. Flowering plants or angiosperms (Angiospermae) are by far the most common, with about 300,000 or so known species. The nonflowering gymnosperms—a term that means "naked seeds"—consist of conifers (Pinopsida), ginkgos (Ginkgoopsida), and cycads and gnetophytes (Cycadopsida). Angiosperms may be further divided into monocots and dicots (also known as eudicots). Monocots have a single leaf when they first emerge from the seed, while dicots have two leaves. Moreover, monocot vascular tissue is distributed throughout the plant's stem, whereas dicots organize theirs into a circle of little bundles. A few species are considered basal angiosperms, which have a mixture of monocot-like and dicot-like traits.

Generally, when people talk about the particulars of plant anatomy, they are primarily referring to those of vascular plants—leaves, stems, and roots, as opposed to the phyllids and rhizoids of nonvascular species. To retain a modicum of simplicity amid the tangled thicket of botanical terminology, I default to words for vascular plant anatomy in this book. For example, I may use "leaves" collectively instead of "leaves and/or phyllids"; even though you understand the basic difference between the two, it is okay to casually refer to the phyllids of a moss as "leaves."

Plants generally have some sort of root system that keeps them anchored to their substrate—whatever they're growing on—and absorbs nutrients and water. The rhizoids of bryophytes are not as complex as the true roots of vascular plants, which may grow in several different configurations depending on the taxa. The aboveground portion of a vascular plant is supported by a stem of some sort, which supports its leaves, flowers, seeds, and fruits; nonvascular plants do not have true stems, so their phyllids and sporangia grow out of the rhizoids.

Leaves are exceptionally crucial to a plant's survival, as the chloroplasts within them manufacture food from the sun's energy through photosynthesis. The leaves and stems are covered in stomata, pores that open or close to allow the plant to breathe, or release or retain moisture. Some plants cover themselves in part or in whole with defensive mechanisms against herbivorous animals, such as sharp thorns or irritating oils; others may carry dangerous toxins in their tissues that can cause anything from a terrible taste to death if eaten.

As mentioned earlier, a plant's sexual reproductive structures can often be crucial to accurate identification. While sometimes differentiation between species may come down to looking at them under a microscope, in most cases your eyes and perhaps a magnifier will suffice. Nonvascular plants grow temporary structures known as sporangia in which spores are formed and then released. Vascular plants (other than the few spore-producing exceptions mentioned earlier) reproduce using eggs contained within ovules and sperm packaged in pollen.

In gymnosperms, ovules and pollen develop in separate female and male cones; when the male cones are mature, they open, and pollen is carried by the wind. Pollen that lands on an open female cone can then release sperm to fertilize the eggs. Angiosperms have more elaborate structures for their eggs and sperm: flowers. While some are also wind-pollinated, many species have bright, aromatic flowers and sweet nectar to entice insects and other animals to visit, who carry pollen from plant to plant that may find a receptive ovule. The pollen develops on a structure called the stamen; it has a stem-like support called the filament, at the end of which is the anther that holds the pollen. The ovules, on the other hand, are nestled deep inside the flower in the ovary, above which extends a narrow tube called the stigma; the ovary and stigma together form the carpel. Pollen needs to land on the stigma for sperm to swim down to the eggs in the ovules. Some plant flowers have both pollen and ovules, while others have only one or the other, so a plant of the opposite sex must be physically close enough for fertilization to occur. There are even plants that can fertilize themselves, creating the next generation even in isolation!

The petals that surround the plant's reproductive organs help recruit animals in making sure the pollen has a direct route to a receptive pistil. Their bright colors grab the attention of potential pollinators, while sweet or otherwise strong scents create additional attraction, especially for flowers that bloom only at night. Glands called nectaries produce the sweet liquid nectar that many pollinators enjoy eating; the nectaries are often located deep enough inside the flower that animals must brush against the anthers or pistil to access their reward. Sepals are leaf-like structures that protect the petals during the flower's development and may act as physical support once it is open.

There are numerous other anatomical features of plants that I have either glossed over or not mentioned at all, but the above are some of the most common and basic that you'll run across as you practice plant identification. Let's get deeper into some of the specific traits you might observe about a plant that may help you figure out what it is.

Color

Plants are literally defined by the color green. The collective plant life of a place, especially in a cultivated area, is known as greenery, while a park or other nature-dominated location full of a variety of plants is termed green space. A building especially for growing temperature-sensitive plants is a greenhouse. And Earth is sometimes known as the green planet because of the significant amounts of green visible on land when viewed from space.

All this green, of course, is because of chlorophyll, the pigment in plants that is integral to photosynthesis; it absorbs red and blue light while reflecting green light, hence appearing green to our eyes. The ancient cyanobacterial ancestors of plants evolved billions of years ago, and they outcompeted other bacteria that used a molecule known as retinal for their photosynthesis. Because retinal absorbs green light and reflects red and blue light, it appears purple. Had these other bacteria gained an edge over the cyanobacteria, we might be living on a purple planet today.

As it is, green reigns supreme on Earth—but that's just where the colors start. First, the plant kingdom is full of an entire spectrum of green shades, from the pale bluish frost of sagebrush (*Artemisia* spp.) to the dark, leathery leaves of kinnikinnick (*Arctostaphylos uva-ursi*), plus any number of artificial cultivars bred for brighter, darker, variegated, or otherwise unusual variations on a species' normal hue.

Get used to describing the particular quality of green you see, because that will help you when trying to convey the coloration of a plant's foliage. In most cases, a plant won't be all one shade; take a closer look at the leaves, stems, trunks, branches, and roots of a given species, and you're likely to find a greater variety of hues. Maybe it's the reddish tint on the twigs and new leaves of an evergreen huckleberry bush (*Vaccinium ovatum*) or the tiny white hairs covering the leaf of hairy cat's ear (*Hypochaeris radicata*). The fresh spring needles of many conifers, such as Sitka spruce, are often lighter green than older foliage. I've pulled up a lot of Scotch broom (*Cytisus scoparius*), which is invasive here in the Pacific Northwest. From a distance, a stand of broom looks like a uniform emerald wall, but on closer inspection, the thinnest stems are often just a little lighter in color than the leaves, while the main stem is covered in grayish-brown bark.

The woody portions of trees and shrubs are usually some shade of brown as well, although this varies. The rainbow eucalyptus (*Eucalyptus deglupta*) is perhaps the most dramatic example, with its bark peeling in layers of green, red, blue, yellow, and orange. Trees whose bark hosts other species may have a multicolored appearance; the reddish, stringy bark of western red cedar (*Thuja plicata*) frequently hosts *Cladonia* lichens, which add splashes of pale green.

But it is the reproductive structures of plants that are often the most impressively colored. Flowers' bright tones attract pollinators, and you can find them in just about every color the human eye can see—as well as some that we can't. Evidence shows that many insect pollinators can see ultraviolet light, and plants use it to create even more impressive visual displays; what we see as a plain yellow dandelion appears to a bee as pale blue with a bright pink spot in the middle, advertising "Nectar Here!"

All of this means that, once again, you'll need to not just get the overall color impression of a plant, but also look closer at the details. Stand a few feet away from the plant to get that first impression of "Well, it's a medium tone of green," and then start counting all the individual shades of green and other colors you notice on leaves, stems, flowers, and so forth. Use your magnifying glass if you like and take some photos that highlight particularly interesting shades or patterns.

Make sure you are really checking everything. It's easy to say, "Well, this bitter cherry (*Prunus emarginata*) flower is white with a yellow center." But if you examine it further, you may notice that the pale-yellow center where the petals emerge is really more of a yellowish green, the anthers are white with little yellow-to-brown balls of pollen on the ends, and the pistil darkens to a golden-brown color at its tip. While these details may not seem immediately important, sometimes the differentiation between two plant species may come down to slight color differences in one feature.

As with animals, if you take some colored pencils, crayons, or other art supplies with you into the field, you can use them to make a visual record of what you see. I find that drawing makes me notice those tiny details more than taking a few quick photos, and I can draw a particular part of the plant—say, an especially colorful leaf—larger than life so that the various hues and markings are more obvious.

Size and Proportions

Plants come in a wide range of sizes. I can look at a cluster of tiny duckweed plants (*Lemna* sp.), each about ¼ inch long, floating at the edge of the lake near my home, and then turn around and almost immediately walk into the trunk of a Sitka spruce towering a hundred feet overhead. And these aren't even the smallest or largest plant species out there.

Height is generally the most important measurement to take when looking at a plant, as this is what field guides and other resources most commonly refer to. Some may also mention the girth of tree trunks or the average width of shrubs, and it's not uncommon to see measurements listed for flowers, fruits, and leaves. So, as with color, get the big picture and then zoom in on the details. Since plants generally tend to stay still, feel free to either use a tape measure or just estimate the length and width of various parts of a specimen, as well as the whole thing.

As with colors, sometimes telling two similar species apart may come down to differences in size and proportion. If someone presented me with the full-grown, ripe berries of both trailing blackberry (*Rubus ursinus*) and Himalayan blackberry (*Rubus armeniacus*), I would know that the smaller of the two would be the trailing species. And while walking along the coastline here in Washington, I can instantly pick out our native American dune grass (*Leymus mollis*) amid a sea of invasive *Ammophila* or *Calamagrostis* grasses because its leaves are much wider, like lengths of blue-green satin ribbon poking up out of the sand.

Shape

Here again, you want both the big picture and the little details. Many large conifers here in the Pacific Northwest, when mature, look like tall, bare trunks with clusters of green needle-covered branches in a vaguely cone-shaped configuration at the top. The needles themselves are skinny and usually pointed at the tip, and the bark of the trunk is roughly textured.

Since that describes almost every native conifer across the region, you're going to need to get far more specific. It can be difficult to assess the exact silhouette of a tall tree from the ground, so home in on what you can see. Look at those needles. Are they long and slender or short and stubby, and how many of them grow out of one spot on the twig? Are the ends sharp or blunt? If you neatly snip a needle in half, look at the cross-section (your magnifying glass may help here) and see what shape it is.

Bark is the best thing to look at when identifying trees and some shrubs. Deciduous species lose their leaves in winter, but bark persists year-round. In addition to the color of the bark, you also want to pay attention to textures and patterns. Many trees

Sitka spruce (*Picea sitchensis*) and
western hemlock (*Tsuga heterophylla*)

have deeply grooved bark, and how those grooves are shaped can be an indicator of species. A member of the Angora Hiking Club in Astoria, Oregon, gave me the best arboreal mnemonic: "potato chips and bacon strips." The former refers to the round patches of bark on the trunk of a Sitka spruce that do look like a crunchy snack only a beaver could love, while the latter describes the long, slender grooves on the trunk of a western hemlock (*Tsuga heterophylla*), like the stripes on a slab of bacon.

Even the smoothest bark still has various irregularities, like scars from old branches, injuries, or diseases. You might see lines or spots of rough, porous bark all over the tree's trunk and branches; these are called lenticels, and they allow air to pass through the otherwise airtight bark. Look closely, as differently colored patches on the tree may actually be colonies of moss or lichens.

Bark color and texture can change dramatically over the lifetime of a tree. A young Douglas fir, for example, has relatively smooth gray bark with brown lenticels. As it grows older, the bark becomes much rougher and patchier with increasingly deep fissures, as well as reddish patches where old bark has flaked away and what's uncovered hasn't weathered to gray just yet. A mature tree may have grooves deep enough for me to fit my entire fist in them with room to spare.

Leaf shapes, arrangements, margins, and veins are very important traits for identifying plants in general, especially those without bark. The shape is the overall outline of the leaf, and each type of shape has its own name; for example, a long, slender leaf with a pointed tip is referred to as lanceolate, while a diamond-shaped leaf is rhomboid. Look at how the leaves are arranged on the stem. If they grow in matched pairs, that's called pinnate; oddly pinnate means there is one leaf alone at the tip of the stem, while evenly pinnate arrangements end with a pair of leaves. If the leaves are unevenly matched, like footsteps "walking" up the stem, that's known as alternate.

The edges of a leaf are known as margins. Some leaves may have smooth margins, known as entire, while serrated leaves

RHOMBOID TO OVATE
SHAPE, SERRATED EDGES,
PINNATE VEINS

OVATE SHAPE, SMOOTH
EDGES, ARCTUATE VEINS

ELLIPTICAL LEAF SHAPE,
TOOTHED EDGES, PINNATE
TO ARCTUATE VEINS

red alder (*Alnus rubra*), black locust (*Robinia pseudoacacia*),
Virginia creeper (*Parthenocissus quinquefolia*)

have a toothed edge. You can also find doubly serrated margins, where the bigger teeth have smaller teeth on their outer edges. If the teeth are rounded instead of pointed, that's a crenate margin, while a leaf margin with sharp, pointy protrusions is—appropriately—spiny.

The veins of leaves have unique patterns as well. Many of us are familiar with a pinnate leaf that has one central large vein and then several smaller ones branching off in pairs, one on either side. If those smaller veins also have even tinier ones connecting them, that's known as cross-venulate. What if you just have several veins running next to one another along the entire length of the leaf, like those on a blade of grass? Unsurprisingly, we call that parallel venation.

Finally, some deciduous trees and shrubs have distinctive leaves and flower buds that can be seen on their twigs even in winter, months before they begin to grow into new leaves. Examining buds and twigs can help with identification, although it can take a little practice to know what you're looking at. Features to examine include the shape and size of each bud, as well as where it's located on the twig; whether there are scars just below the bud where last year's leaf grew and then was shed; the shape of the twig (straight, crooked, and so on); and whether the twig has lenticels and, if so, what they look like and how they're arranged.

Roots

In most cases, you aren't going to get a good look at the root system of a plant because it's buried underground. You might see the upper surfaces of a tree's roots, especially if there's been a lot of erosion around them, but the only way to see them in their entirety is if the tree gets blown over. However, there may be cases where the roots are accessible, such as when you're weeding your garden, helping to remove invasive species, or planting bare-root plants. I don't recommend just uprooting any plant you find; even if you put it back, the trauma may be too much for it to recover from.

If you do get a chance to examine a plant's roots, notice how they're laid out. Is there a main taproot with smaller rootlets growing off it? Or is there a whole network of roots bristling underneath the plant? How many times do roots branch off into smaller ones? What color are they, and how long are they in relation to the rest of the plant?

Sometimes the roots can give you a hint about the plant's lifestyle. Generally speaking, plants that live in drier locations have longer roots than those in wetter places, allowing them to draw up water from deep in the ground. A plant with a particularly hefty taproot may be storing food in there, like the carrots or sweet potatoes you see at the grocery store. While grasses don't have taproots, their vast root systems can store food for the future; the native grasses of America's Great Plains, for example, have roots that are often multiple times longer than the aboveground portions of the plants and that hold enough energy reserves for them to regrow after being grazed or burned.

Reproductive Structures

Flowers, seeds, spores, and other reproductive structures are among the most distinctive portions of a plant when present. The shape is important, but so is the type of reproductive structure the plant has in the first place. This is why I went into some detail earlier in this chapter about how plants are categorized as vascular versus nonvascular and angiosperm versus gymnosperm. Being able to place a plant within the appropriate categories can be a great start to getting a solid identification, although you may not immediately recognize what you're looking at because there's so much variety.

For example, the sporangia of mosses and liverworts commonly look like tiny capsules on thin stalks growing up above the main layer of phyllids. Fern sporangia appear in sori, which are tiny fuzzy dots on the undersides of the leaves, while horsetails produce strobili—stalks with club-shaped tips full of spores—before their leaves even start to emerge for the year.

Gymnosperms produce via strobili, too. The cones of conifers are the most commonly seen, but cycads also have massive strobili that resemble oversized fir cones. *Ginkgo biloba* reproduces with male cones and female ovules, while some species of gnetophytes have reproductive organs that look like little flowers made of yellow-green bubbles, tiny fruit, or even caterpillars standing on their heads—and all of them are, once again, strobili.

The flowering plants have the most vibrant variety when it comes to reproduction. Flowers range from the tiny, round, green-white blooms of *Amborella trichopoda* to the massive red-and-white five-petaled flower of the stinking corpse lily (*Rafflesia arnoldii*), which can be almost four feet across and smells like rotting meat to attract pollinating flies. They may be shaped like discs, tubes, cups, or funnels; orchids (Orchidaceae) have evolved into a stunning array of exotic shapes, including one that looks like a tiny naked man.

Once ovules are fertilized, the resulting seed grows within a fruit. While we most often think of edible fruits like apples, oranges, berries, and so forth, nuts, grains, and the hulls of sunflower seeds are technically fruits, too. Just like flowers, they may take a wide range of forms, from an avocado that protects one large seed to the raspberry that is composed of dozens of tiny fruits clustered together in one aggregate mass. Roses have a fruit within a fruit; to reach the seeds, you first must split open the round, fleshy rose hip and then the small, dry brown achenes, each of which protects a single seed.

While all these reproductive structures may have distinctive appearances that help greatly in plant identification, the downside is that, for the most part, they aren't present year-round, since plants have particular times of the year in which they are fertile. Once that season has passed, you may have to make do with dried-up leftovers, if there even are any. Identification is easy if

Gymnosperm

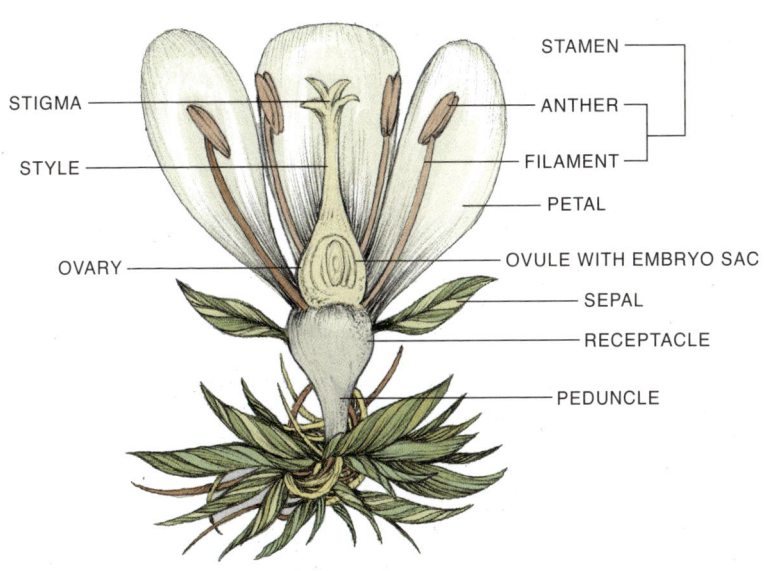

CONE SCALE

CONE SCALE

IMMATURE
SEEDS

RIPENED
SEEDS

CLOSED CONE

OPEN CONE

STAMEN

STIGMA

ANTHER

STYLE

FILAMENT

PETAL

OVARY

OVULE WITH EMBRYO SAC

SEPAL

RECEPTACLE

PEDUNCLE

Angiosperm

you find last year's old cones on the ground around a solitary pine tree, but if several conifer species are growing together, it may be unclear which fallen cones came from which tree. Berries are useful only until they're either eaten by wildlife or dry up and decay. Flowers are even more ephemeral, sometimes lasting mere days, and you might not recognize the sporangia of bryophytes or ferns if you don't know what you're looking for.

The good news is that the more nature identification you practice, the more you'll notice the seasonal cycles of the plants you're observing. From about late February to the end of October, I can tell what time of the year it is where I am simply by which plants are in bloom (and whose pollen is making my sinuses stuffy!). Certain times of the year are especially good for finding particular reproductive stages; spring is prime wildflower time here, but I have to wait until about August before I can do any serious berry picking. Not that all the plants stick to that schedule, of course; by the first of June, I'm already eating salmonberries (*Rubus spectabilis*), and in October I still get to enjoy the blooms of Douglas's aster (*Symphyotrichum subspicatum*).

So, after all that, what should you be looking for when trying to use reproductive structures in plant identification? Here are a few tips:

Numbers: Often families or other groups of plants have consistent numbers of certain parts, such as petals and sepals, anthers, seeds, and so on; this is known as the plant's merosity. Some plants are isomerous, having the same number of sepals, petals, anthers, and carpels in one flower, while the numbers vary in anisomerous species.

Shapes: Describe the flower, fruit, or spores in whatever words you prefer; you can use technical terminology like infundibuliform versus hypocrateriform, but it's just as valid to say a flower is funnel-shaped versus tube-shaped. Note whether a structure is symmetrical or not; could you cut the flower or fruit in half and have both halves be mirror images of each other? If, like me, you occasionally forget words and you can't remember the term *sori* when looking at a fern, it is perfectly acceptable to describe them in your notes as "spore buttons" and then go look up the proper name later.

Season: We'll talk more about phenology—the study of seasonal cycles in plants and other organisms—later in this chapter, but it's important to note what time of year you're observing the reproductive structures and what they look like right now, as well as any time you see that species again in the future. Since the flowers and fruits of a given species may look different from each other, try to familiarize yourself with all the stages of the species you're trying to identify. Ideally, you want to eventually know enough about a plant species that you can recognize it year-round, whether it has fresh flowers, or old, dry seeds, or something in between, or none of the above.

Colors: I want to reiterate what I said earlier about colors because they can be important in gauging the reproductive stage of a plant. I can tell when streambank lupine (*Lupinus rivularis*) is in bloom a hundred yards away because of the vivid purple of its flowers. As the fertilized seed pods mature and then dry, they change from light green to grayish brown. Thimbleberry (*Rubus parviflorus*) fruits reach the perfect ripeness (in my opinion, anyway) when they are a bright cardinal red, but before they deepen to a raspberry shade.

Texture

Visual and tactile textures are worth noting, although touching a plant with your bare hands can be risky if you don't know what it is. Start with the visual texture and be descriptive about it, such as "green, glossy leaves that look like they were coated in wax" or "stringy red bark that reminded me a little bit of shredded corned beef." You may notice that the texture changes depending on whether the plant is wet or not, especially if it has gotten dehydrated during a dry spell; changes in life stages can also affect its visual texture, such as deciduous leaves that are about to fall.

Tactile texture needs to be handled more carefully—literally! Some plants' saps or oils can cause nasty allergic reactions, and thorns are no fun when they get embedded in your skin. Before you go out, familiarize yourself with any poisonous plants in your area so you're more likely to recognize them before you even try handling them. Carry some rubber medical gloves in your backpack if you plan to touch unknown plants; then remove them by turning them inside out so you don't accidentally touch whatever substances got on them, and then dispose of them in the trash.

Tactile texture often matches the visual texture. The leaves of great mullein (*Verbascum thapsus*) both look and feel fuzzy, while the foliage of Pacific wax myrtle (*Morella californica*) is, in fact, waxy in appearance and touch. Large thorns, like those on acacia trees (*Acacia* sp.), are hard to miss. Others may be smaller and tough to see if you give the plant only a casual glance; I've learned the hard way that the tiny prickles of trailing blackberry carry a nasty sting in spite of their size. Cacti famously are covered in spines, but don't be fooled by those that look soft and touchable, as their fine hair-like filaments can still cause serious skin irritation. Stinging nettle's (*Urtica dioica*) leaves may look a little hairy, but those hairs release irritating chemicals like histamines that can cause a burning rash.

Should you accidentally brush up against a plant that causes you instant regret in the form of burning or other irritation, make note of that. If you come home with contact dermatitis that appears

Some plant defenses are quite dangerous to humans. Numerous species in the carrot family (Apiaceae), citrus family (Rutaceae), and mulberry family (Moraceae) can cause phytophotodermatitis. If you get their sap on your skin, it makes you much more sensitive to sunlight, and you may quickly develop nasty burns and blisters. All parts of the manchineel tree (*Hippomane mancinella*), found from southern North America to northern South America, can cause very serious contact dermatitis, as can rain dripping through its foliage, and should it get in your eyes, it can cause at least temporary blindness. The species you most want to avoid touching, though, is the gympie-gympie (*Dendrocnide moroides*), found in Australia and Malesia; it is covered in fine hairs whose hypodermic action injects an incredibly painful toxin that causes severe symptoms lasting weeks or months. The hairs are so fine that they may break off the plant and float into the air, and if inhaled can cause serious, even chronic, respiratory problems.

hours after contact with any plant, that's also something to record. Not only might this help you narrow down what species you literally ran into, but it'll make it more likely you can avoid touching them in the future.

Smell and Taste

Just about everyone has smelled flowers at one point or another, and fruits often have sweet or otherwise notable aromas, too. But other parts of a plant may have distinct scents. When I head east of the Cascades, I always like to crush a few sagebrush leaves to smell their pungent oils. Walk through a grove of ponderosa pines (*Pinus ponderosa*) on a hot day, and your nose may catch a whiff of vanilla emanating from their bark. Any time you observe a plant with a notable smell, make note of it and what part it came from, if possible.

Be reasonably cautious when smelling flowers, just in case you end up being sensitive to even small amounts of pollen. It's much riskier to try tasting a plant you don't know for sure is edible. Even if you don't ingest a significant enough amount to be properly poisoned, you can still end up with localized irritation in your mouth, in all its burning, painful, blisters-on-mucous-membranes glory. Moreover, there are enough other ways to positively identify a plant that chewing on it isn't necessary.

Habitat

Unlike animals, plants are stuck in one place from the moment the seeds land on a suitable substrate. In fact, most seeds never get a chance to fully germinate, since they get digested, destroyed, or end up in too harsh a spot. So, when you see a plant growing successfully, that says that there's something about its location that it needs to thrive. Even if you don't know the entire story of how this plant interacts with its environment, record as much information as you can about where you found it. Field guides and other identification resources often mention preferred habitats of the species they list, and once you've positively identified the plant species, you can learn more about why it likes to grow in these places.

Pay attention to multiple factors when considering habitats. First, look at the overall ecosystem: Are you in a forest or meadow, wetland or desert, or some other place entirely? How abundant and diverse are the other living beings you see here, including other plants? Look at how much sun exposure this particular plant is getting and whether larger plants, landforms, or other obstacles get in the way of the sunlight for some or all of the day. When bad weather comes through, how protected might this plant be? Even a small shelter from the wind can give some plants a necessary edge for survival. Some plants must deal with other physical challenges, like being trampled by animals, covered by sand dunes or mudslides, or periodically inundated by floods. Water, of course, is paramount to any plant's day-to-day life, so consider how much rain and other moisture this one may get, and whether wind and sun evaporate it quickly or shade helps preserve it.

If one force dictates the lives of all but the most parasitic plants, it is sunlight. Any gardener will tell you that each species has its preferences; some prefer full light, while others have evolved in shadier areas and shrink back into more sheltered places. Take note of how much light the plant you're observing is getting as you're looking at it, and then predict whether that may change throughout the day as the sun moves across the sky. Some plants may prefer more sun but be stuck in the shade of their larger neighbors; a young sapling might spend decades eking out whatever scraps of light it can get beneath the branches of its elders before an older tree dies and an opening in the canopy finally appears.

You don't have to be a soil scientist to learn a bit about the substrate in which the plant is growing. If you can dig in the dirt around it, notice whether it seems to be especially crumbly or sandy, or if it sticks together with clay. Pay attention to how much organic material, such as living and dead plant matter, overlays the soil, and whether you see insects, worms, and other small animals in it. If the soil is dry, pour some water on it, let it sit a few minutes, and then see how hydration changes the consistency. Does it become saturated quickly, or does water seem to flow off the soil's surface? Look around for any rocks or gravel in the soil that may be obstacles to plant roots. Notice the soil's color; a light gray soil similar in color to nearby rocks might indicate that it has a high mineral content, while a darker brown soil is likely to have more organic material, although these are, of course, generalizations.

Life Cycles

I already mentioned the importance of phenology in regard to a plant's reproductive organs, but it's a crucial part of understanding the rest of a plant's biology, too. First, plants have varying life cycles. Annuals live a year or less and, in some cases, may have multiple generations a year; biennials live two years and usually reproduce in their second year; and perennials may live for many years. In temperate areas, plant communities tend to go dormant in fall through winter, and then "wake up" in spring and stay active throughout summer. How an individual plant responds to seasonal changes varies. Some perennials and biennials persist through winter, even retaining their foliage, while others drop their leaves or their entire aboveground portion; and annuals die out entirely, and the next generation germinates the following spring.

Chances are that the diversity of plants you see in summer will be much greater than what's apparent in winter. May and June are my favorite times for plant identification in the Pacific Northwest because everything has come back from winter dormancy, seeds have sprouted into easily distinguishable plants, and flowers are blooming all over the place. However, because our winters are comparatively mild, we have a significant number of evergreen plants besides the conifers. Not only are they easy to spot as green patches in otherwise brown winter woods, but the list of plants that keep their foliage year-round is relatively short, meaning identification is often faster.

When you encounter a plant for the first time, you may not immediately know how long it normally lives or where it is, phenologically speaking. A combination of physical traits, such as those reproductive structures and paying attention to what time of year it is, can offer some insight. Let's say I see a small pearly everlasting (*Anaphalis margaritacea*) growing amid dune grass in May, but I don't know what it is. It might be a young plant, but it could also be one that never gets more than a few inches tall;

without positive identification, I can't say for sure. But if I return in August and find that it is much taller and has flowers, I now understand that it was just getting started when I first saw it a few months before. I could come back and visit it the following May and see if it is growing in the exact same spot, which would indicate a biennial or perennial.

Of course, if I managed to positively identify it as pearly everlasting the first time I saw it based on traits like the shape and color of its leaves, I would be able to get the phenological information that much more quickly by reading up on the species. It's still helpful to witness these stages and cycles in person whenever possible; I find it makes them stick in my memory better than just reading about them in a book or on a website because I'm observing them firsthand.

This is why it's important to always date your observations when you record them. You may get only a snapshot of that plant's life cycle, but it could be the key to its identity. If the phenology of the species you're researching matches that of the plant you've observed, that's a point in favor of it being a positive identification.

A Couple More Things

When you take pictures of plants, it's important to get an overall view of the whole thing, and then take some shots of details like leaves, flowers, stems, and the like. Make sure you photograph anything that seems especially notable, like unusual color patterns, textures, and so forth. The more pictures you can get of your specimen, the better a record you'll have to look back on when trying to identify it.

It can be challenging to get a good picture of a plant that's growing amid a whole bunch of others. Sometimes it blends in visually; the camera might also try to focus on other nearby plants instead of the one you want. The blank inside cover or back page of a field guide or journal makes a great background to stick right behind the plant so it's isolated. Sometimes, especially if you're dealing with limited equipment like a basic phone camera, it's going to be difficult to convey the true depth or height of a plant, especially a large one like a tree, so do the best you can with what you've got, and supplement your photos with extensive, detailed written or voice-recorded notes.

If you see a particular plant away from an established trail, you need to be very, very careful not to damage the surrounding habitat if you're going in for a closer look. That means don't trample or otherwise damage other plants, fungi, or animals along the way, and avoid rolling over rocks and logs or creating divots in the ground that could lead to erosion over time. Be gentle with the plant you're observing, too; if you're pulling a branch down closer where you can see it, don't break it or pull off leaves or flowers. And while it may be tempting to take a few blooms home to put in a vase on your kitchen table, resist the urge. Some plants are quite fragile; as one example, picking a trillium (*Trillium* sp.) can kill the entire rhizome even if you don't uproot it, as the loss of all its leaves starves the plant.

If you're looking for other cooperative subjects for your photography, fungi are similarly stationary and offer even more opportunities for nature identification. The next chapter explores this fascinating and unusual kingdom of beings.

PLANT IDENTIFICATION TEMPLATE

Species Observed:
Stag's-horn clubmoss (*Lycopodium clavatum*)

What, When, and Where?

DATE: July 1, 2019

LOCATION: Coastal Forest Loop Trail, Cape Disappointment State Park, Ilwaco, WA

NUMBER OBSERVED: >100 individual plants

ALIVE OR DEAD? Alive

VASCULAR OR NONVASCULAR? Vascular

SEEDS OR SPORES? Spores

ANGIOSPERM OR GYMNOSPERM? N/A

HABITAT/LIGHT: Growing on the floor of a mature mixed conifer forest in soil fed by decaying plant matter. Partial sun due to canopy openings created by the trail.

Leaf Details

SHAPE/ARRANGEMENT: Lance-shaped, slender, spiral arrangement

MARGINS: Smooth

VEINS: One single, central vein

What's It Look Like?

SIZE/PROPORTIONS: Individual plants could be over 2 feet long. Slender stem—¼-inch diameter packed with many microphylls not exceeding ¼ inch long. Reproductive structures consist of up to 6-inch-tall slender bare stalks with strobili—1 inch of the total length.

OVERALL SHAPE/DETAIL SHAPES: Each plant made of several stems branching multiple times, looking like a large bed of moss but with more structure. Stems covered with many tiny, tightly packed leaves, giving them a "skinny caterpillar" appearance. Leaves are very slender and lance-shaped with pointed tips.

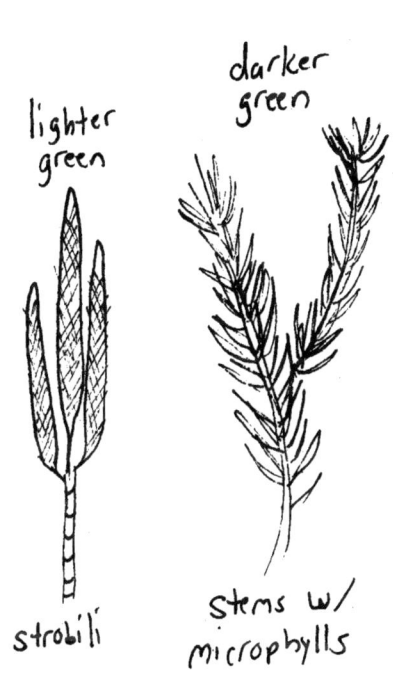

Strobili stalks are segmented or ribbed; strobili slender and tightly scaled, reminiscent of an unopened conifer cone.

COLORS/PATTERNS: First impression is a mass of medium green. On closer inspection, strobili and stalks are lighter yellow-green. Stems lighter than leaves.

VISUAL TEXTURE: Appears fuzzy but looks like leaves might be stiffer and more prickly than those of moss.

TACTILE TEXTURE: Soft, although not quite as soft as moss due to stems. Not prickly.

LIFE CYCLE: Likely perennial, but need to check back in winter.

CASE STUDY
Species: Stag's-horn clubmoss (*Lycopodium clavatum*)

DATE: July 1, 2019

LOCATION: Cape Disappointment State Park, Ilwaco, WA

OBSERVATION NOTES: I've hiked the 1½-mile Coast Forest Loop Trail at Cape Disappointment many times; one of the features of this mature western hemlock–Sitka spruce forest is its incredible biodiversity. In the northeast section of the loop, there is a large patch of moss-like plants with tendrils that may reach up several inches from the ground. After passing it several times, I finally decided to figure out what it was.

Initial impression was of a large mass of fuzzy, medium green carpeting the ground along the trail where openings in the canopy allowed some, but not full, exposure to sunlight. Each plant, which could be more than 2 feet long, had several stems that branched multiple times, and all were covered in numerous tightly packed, slender, lance-shaped microphylls (leaves). The microphylls were generally less than ¼ inch long, tightly packed, and arranged in spirals around the stems.

The plants looked rather soft and fuzzy, similar to moss. Their tactile texture matched this and were much less prickly than I had originally anticipated. Because of the thickness of their stems—about ¼ inch in diameter—they had more physical substance to them than moss. They were growing in rich, damp soil with no rocks or sand evident, and were likely fertilized by decaying plant matter on the forest floor.

Some of the plants had produced tall stalks with strobili on them pointing toward the sky. The stalks were bare of microphylls, with a segmented or ribbed appearance and were a lighter yellowish green than the rest of the plant. They averaged 4 to 6 inches in height, about an inch or so of which consisted of the strobili themselves at the tops of the stalks. Each strobilus was slender with a tightly scaled appearance, similar to the pattern found on an unopened conifer cone.

DISCUSSION: Clubmosses are so named because they look like patches of moss that produce club-like strobili. And it would be tempting to just assume this is a moss due to the similarity of their microphylls. However, remember that true mosses are nonvascular bryophytes, and, therefore, they would neither have stems nor would they reach the lengths of the observed plants. Mosses also do not produce strobili; instead, they have much shorter, simpler capsule-like sporophytes.

Once I had determined that this was a type of clubmoss and not a true moss, that narrowed my options down quite a bit. The most common clubmoss in Pacific County is the stag's-horn clubmoss, which is a good fit visually. About the only other species of Lycopodiopsida commonly found in the area is Oregon spike-moss (*Selaginella oregana*). However, that plant is more often found growing as an epiphyte on trees or rocks. The plant is markedly more slender than stag's-horn clubmoss, and its microphylls are tinier and not as tightly packed on the stem, giving it a spikier visual texture.

Because I have access to this patch of stag's-horn clubmoss year-round, I can check back on it throughout the year. Once the strobili are mature, usually by early fall, they turn yellow and gain more texture as the "scales" open to allow the spores to spread. This plant is perennial and persists throughout winter rather than dying back in seasonal senescence.

PLANT IDENTIFICATION TEMPLATE

Species Observed:
Hairy cat's ear (*Hypochaeris radicata*)

What, When, and Where?

DATE: May 19, 2017

LOCATION: Home, Long Beach, WA, front yard

NUMBER OBSERVED: >200 individual plants (estimated)

ALIVE OR DEAD? Alive

VASCULAR OR NONVASCULAR? Vascular

SEEDS OR SPORES? Seeds

ANGIOSPERM OR GYMNOSPERM? Angiosperm

HABITAT/LIGHT: Growing on unmowed lawn in full sunlight, but it seems to like any place that isn't shady.

Leaf Details

SHAPE/ARRANGEMENT: Rosette at base of the plant

MARGINS: Pinnatisect with rounded, opposite lobes

VEINS: Appears pinnate but curving toward tip of leaf; possible very faint cross-venulation?

What's It Look Like?

SIZE/PROPORTIONS: Tall flowering plants with slender stems that may approach 18 inches tall in exceptional specimens. Flowers average 1 inch in diameter, leaves are slender <3 inches wide and up to 8 inches long.

OVERALL SHAPE/DETAIL SHAPES: First impression is of a tall, gangly dandelion (*Taraxacum officinale*). However, the most obvious difference is that the stems are solid, not fleshy and hollow, and the main stem may branch several times with a flower at the end of each branch. The leaves are pinnatisect, but the lobes are more rounded than on dandelions and have fine white hairs all over them. Leaves are arranged in a rosette flat on the ground, with stems growing from center.

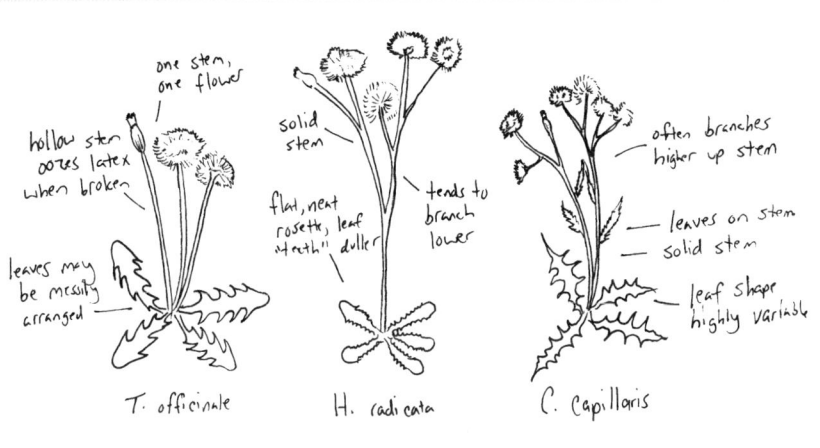

COLORS/PATTERNS: Medium- to dark-green leaves and stems with solid color, but leaves have a layer of white hairs that can give them a frosted appearance. Flowers are bright yellow, turning to dark brown seeds with fluffy white parachutes.

VISUAL TEXTURE: Leaves have a decidedly fuzzy or hairy appearance; flowers look soft to the touch, stems firm.

TACTILE TEXTURE: The tactile matches the visual pretty closely. Stem is flexible but may be stiff enough closer to base to snap.

LIFE CYCLE: Probable annual given similarity to dandelions; all visible plants have mature flowers and flower buds.

CASE STUDY

Species: Hairy cat's ear (*Hypochaeris radicata*)

DATE: May 19, 2017

LOCATION: Home, Long Beach, WA, front yard

OBSERVATION NOTES: This prolific plant looks superficially like a common dandelion (*Taraxacum officinale*). However, rather than a single fleshy, hollow stem per flower, it has several slender green stems, which may branch multiple times, and each branch has a flower at the end of it. The stems are solid and covered in tiny white hairs.

At the base of the plant is a rosette of leaves. Like those of the dandelion, they are pinnatisect—long and slender, with opposite pairs of lobes along the margins. However, the lobes here are much more rounded than the pointed lobes of dandelions. There's a central vein with several smaller veins branching off; some of them terminate into their neighbors, and others curve toward the end of

the leaf. Like the stems, the leaves also have a healthy coating of white hairs all over.

The flowers are almost exactly like those of dandelions, composed of dozens of tiny yellow petals arranged in concentric circles and several yellow anthers as well. However, they are just a little smaller, as are the seed heads. Each seed is small, slender, and dark brown, and it has a fluffy white parachute to catch the wind.

These plants aren't especially picky where they grow, so long as they have full sun. They grow in both damp meadows and dry dunes and will happily colonize disturbed areas; this suggests they are likely annuals that have evolved to take advantage of any open area. Previous casual observation shows that these may persist far into autumn and show up early in spring, if not the tail end of winter.

DISCUSSION: The first time I really took a look at a hairy cat's ear, I assumed it was a sort of hawksbeard (*Crepis* sp.). Many other people think it's a dandelion (*Taraxacum* sp.). And botanists agree that all the above are examples of those frustrating damned yellow composites (DYC) that I talked about back in Chapter 4.

Whenever you have "something that looks like a dandelion," you can't look just at the flowers. Instead, start with the stems. How are they structured? While a common dandelion has one flower per stem, the stems of hairy cat's ear branch and then have a flower on each branch. Those stems are also solid, whereas dandelion stems are hollow and fleshy in texture and leak a white latex when damaged.

Both species have a rosette of leaves at the base of the plant. Hairy cat's ear leaves usually lie flat against the ground, whereas the dandelion's may be a little more unruly. The lobes of dandelion leaves are pointier and often turn back toward the center of the rosette. And they lack the characteristic white fuzz that gives hairy cat's ear its common name.

A couple of other look-alikes grow in the area. The smooth hawksbeard (*Crepis capillaris*) that I thought I had at first also has branching stems, but they branch much higher up, and they often have a few unusual lance-shaped leaves with toothed margins growing on the stem between the basal rosette on the ground and the flowering branches. The lobes on its pinnatisect leaves are also pointier than on a hairy cat's ear.

Prickly sow-thistle (*Sonchus asper*) does have those dandelion-like flowers. However, they grow in a tiny cluster at the top of a single thick stem that can be 6 feet or taller. It may start as a messy rosette of leaves on the ground, but soon a single thick stem appears, with more pinnatisect leaves growing in an alternate pattern all the way up. A tiny little rosette of leaves grows just below the flowers, making it look like a miniature dandelion growing at the top.

CHAPTER 7

How to Identify Fungi

chicken of the woods (*Laetiporus sulphureus*)

OF ALL THE ORGANISMS THAT I HELP PEOPLE IDENTIFY, I get the most requests for and questions about fungi. Mushroom hunting has become a hot(ter) topic in recent years, and the pandemic brought a surge of interest in foraging for wild edible plants and fungi. Thankfully, I'm also a mushroom enthusiast fortunate enough to live in an area of the Pacific Northwest with some amazing fungal biodiversity, so I've had ample opportunity to identify species almost year-round.

Because mushrooms sprout out of the ground and don't tend to move around much, they're often mistaken for plants. As it turns out, fungi are more closely related to us animals. The way in which our animal cells and fungal cells metabolize molecules is so similar, in fact, that pharmaceutical companies developing new antifungal medications must take great care to make sure these chemicals don't harm us as much as the target fungi.

That said, plenty of traits make fungi unique. Let's take a look at the basic anatomy of the mushrooms that pop out of your lawn, rotting logs, or compost piles.

Basic Mushroom Anatomy

A mushroom is the fruiting body of a fungus; that is to say, it is the fungus's temporary reproductive structure. The main body of the fungus is known as the mycelium, which is made of little filaments known as hyphae. The mycelium permeates, or grows throughout, whatever substrate the fungus prefers. Most fungi have a particular preference for soil or decaying wood; some may grow in either one, while others may be found in compost, manure, or those forgotten leftovers in the back of the fridge that everyone's been afraid to look at for weeks.

Savvy mushroom hunters know that if they find a particular mushroom in a given place one year, it's worth going back to check

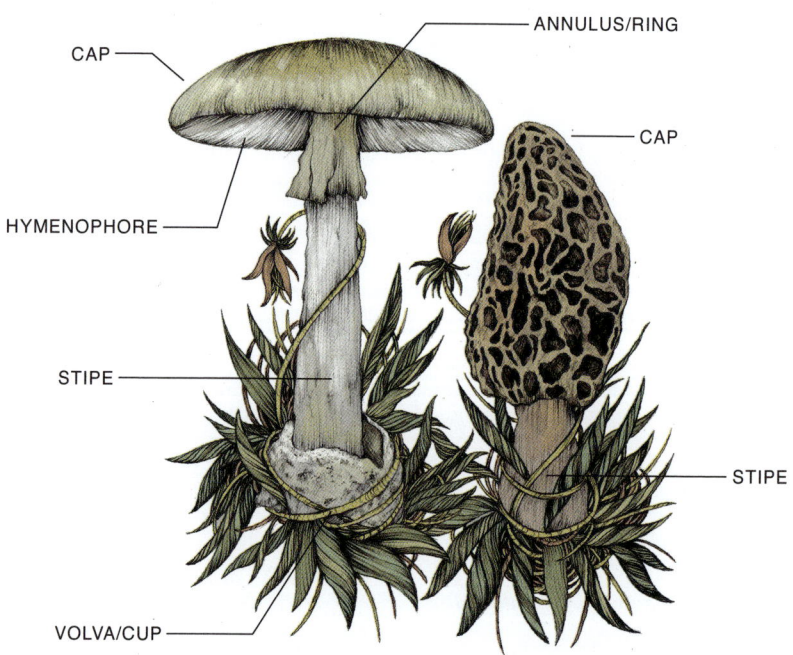

ANNULUS/RING

CAP

CAP

HYMENOPHORE

STIPE

STIPE

VOLVA/CUP

Typical gilled mushroom and morel ascocarp

that spot next year because the mycelium persists year in and year out. It may expand to new parts of its substrate in the ongoing hunt for nutrients, but it and its descendants may colonize an area for decades if conditions are right.

The mushroom itself, on the other hand, is ephemeral. It will last anywhere from a few days to a couple of weeks for most species, spread its spores, and then rot away. Other than a few longer-lasting exceptions, mushrooms are temporary—but mycelium is forever. The mushroom is the most visible part of the fungus, so let's look at its basic structure.

Cap

The cap—also known as the pileus—is the most obvious part of most mushrooms. Many mushrooms look something like open umbrellas; the cap may be flat, domed, convex, or even cone-shaped. Some species even become concave, or bowl-shaped, with age. In basidiomycete mushrooms, the outer edge of the cap is not attached to the stipe (stem), although the ascomycete morels (*Morchella* spp.) are a good example of what I like to call "closed umbrella" mushrooms, since the bottom edge of the cap is attached to the stipe all the way around.

Hymenophore

Underneath the cap, most mushrooms have visible structures that spread spores, known as the hymenophore. Many have gills (known as lamellae), thin sheets of fungal tissue that grow close together and look like the pages in a book. Chanterelles (*Cantharellus* spp.) are an example of mushrooms with false gills; they are thicker than true gills, often more widely spaced and with shallower grooves between them. Whereas true gills look like paper, false gills appear as though they were carved out of wax. Both true and false gills may be decurrent in some species, meaning that they extend beyond the underside of the cap and partway down the stipe.

Some mushrooms have soft, spongy pores, sometimes referred to as tubes; the family Boletaceae is a great example of these porous mushrooms. Most species have pores only under the cap itself, although some have decurrent pores that extend onto the stipe.

Gills and pores aren't the only options; for example, check out the underside of toothed mushrooms like hedgehogs (*Hydnum* spp.), and you'll see little structures that look like icicles or spines. Not every mushroom may have easily visible hymenophores; puffballs (*Lycoperdon, Calvatia,* and other genera) are essentially balls of fungal tissue that release the spores inside only when they pop open at maturity.

Spores

To the naked eye, spores look like very fine colored dust. Under a microscope, they may be various shapes, such as round or elliptical, and with a high enough magnification, you might see texturing on the surface of some of them. You don't need to get quite that close a look to identify more distinctive mushroom species; generally, the spore color will be a good enough clue. I'll talk more about spore prints a little later in this chapter.

You may read about fungi being either basidiomycetes or ascomycetes. They are categorized according to the shape of the tiny structures that produce their spores; you won't be able to see them without a microscope, of course, but it's good to know the difference between the two groups. Basidiomycetes' spore-producing structures are like stalks or clubs with the spores on the ends; you can remember that *ba*sidiomycetes are like *ba*seball *ba*ts. The spore-producing structures of ascomycetes resemble little bags or pouches; therefore, the *asc*omycetes are like *sac*s. (Remember this refers to the microscopic structures, not the entire mushrooms you can see with your naked eye, which may appear in a number of diverse forms.)

There is a third group known as fungi imperfecti or deutero-mycetes. This simply refers to species that can't be easily shuffled into basidiomycete or ascomycete camps. Some don't have fruiting bodies, while others lack spores altogether. Again, you won't know a fungi imperfecti just by looking at it, but now you know what the term means.

Stipe

The stipe is the stem or stalk of the mushroom. It supports the cap and holds it up higher so that its spores may be more easily spread by the wind. Not every mushroom has a distinct stipe; for example, some oyster mushrooms (*Pleurotus* spp.) are laterally attached, consisting of a cap directly attached to the wood out of which it grows. Some stipes are smooth, while others feature unique textures or color patterns. Stipes may also have other structures on them, like the ring and cup examples described below.

Ring

The ring, also known as the annulus or veil, is a band of tissue around the stipe a little below the cap. It is the remnant of a protective membrane that covered the hymenophore while the growing mushroom was pushing its way out of its substrate. Not every mushroom has a ring on its stipe; some never had one, while others had their rings fall off at some point. If you see a ring, take note of it, as it can be an important detail in determining identification.

Cup

Also known as the volva, the cup is also the remainder of a protective membrane. This one covered the entire mushroom as it was developing, until it was big enough to push its way through. The cup often looks like a broken-open eggshell at the base of the stipe. Again, not every mushroom has a cup, and this structure may also become damaged or be missing, so its absence should not necessarily rule out mushrooms that commonly have cups.

Alternate forms of fruiting bodies that fungi may produce include but are not limited to conks, which are shelf-shaped; artist's conk (*Ganoderma applanatum*) is a common example found growing on dead or dying trees worldwide. Jelly fungi, such as *Dacrymyces* spp. or *Tremella mesenterica,* look like little dabs of jam sitting on logs or stumps. Cup fungi resemble fleshy bowls or other vessels poking out of the dirt; orange peel fungus (*Aleuria aurantia*) is one of my favorites, as it literally does look like someone dropped a bit of rind on the ground. For brevity's sake, I will be referring to all fruiting bodies as mushrooms.

Other more specific details of mushroom anatomy exist with their own terms that you may run across now and then in mycological literature, but these ones covered above will be sufficient for using most beginner-friendly field guides and other resources. Now, let's get into some of the other traits you'll want to look for when trying to identify a mushroom for the first time.

Color

The color of the cap will probably be what jumps out at you first because it's the biggest, most obvious portion of the mushroom. Take a close look at the top of the cap to see if it's truly one solid color or whether there are patterns, striations, faded portions, and so on. Check the underside to see what colors you find on the hymenophore and stipe. Make a note of any unique color patterns, such as the red hues found on the white stipe of *Russula sanguinaria* that differentiate it from other *Russula* species with red caps and white stipes.

It's also helpful to cut the mushroom in half to see if the interior has any unique colors. For example, the Pacific golden chanterelle (*Cantharellus formosus*) is golden yellow on the outside and often paler inside. Letting the cut mushroom sit for a few minutes can also help you determine whether the mushroom bruises or not (more on that in a moment).

Pay attention to how developed the mushroom appears, and whether you see mushrooms of different colors but otherwise similar appearances in the same patch. As I've discussed, some species may be a different color when young than when they become older; a very young death cap (*Amanita phalloides*) is white to ivory in color, although with age, the top of the cap often takes on a yellowish or greenish tint. Some death caps may remain white throughout their lifespan, though, which is a good reminder that there can be color variation within the same species. This is why it's exceedingly important to look at multiple photos of a species that you're comparing to the mushroom you have in hand, so you can see whether the color is uniform throughout all members of that species.

Size and Proportions

Make note of the size range of the mushroom species you're trying to identify if you're able to find multiple specimens. Most field guides give the average size range of fully mature mushrooms, so don't count the size of those that are obviously underdeveloped, as I've just discussed. While outliers exist at either end of the spectrum, most individuals of that species should fall within the typical size range. Pay special attention to the diameter of the cap, the diameter of the stipe, and the total height of the mushroom.

Proportions are also important to note. The king bolete has a thick stipe, often with a bulbous end. Some other *Boletus* species also have similar proportions, but if you find a porous mushroom with a slender stipe, it's probably not going to be the king. That said, individual variations are common. Not every king bolete will have a particularly large bulb, but it can still be positively identified by other traits, like a reticulated brown-and-white pattern on the stipe or a cap that looks something like a bread roll.

Shape

There are two levels of assessing the shape of a mushroom: the overall big picture and the little details. The former is your initial impression of the whole mushroom. Is it shaped like an open or closed umbrella? Or a shelf sticking out of a tree? Or a little blob of gel? Maybe you've found a lobster mushroom, which is a *Russula* or *Lactarius* mushroom that has been infected by *Hypomyces lactifluorum* and warped into all sorts of strange shapes reminiscent of alien fungi in a sci-fi movie. Remember that, like color, a shape can change with age.

Next, it's time to check out the little details. This is when you might start noticing that the edges of that yellow-capped mushroom look a little like crocheted lace, a good point in favor of it being *Stropharia ambigua*. Or if you're trying to figure out which species of *Hericium* this giant mass of white fungal "fur" you've found is, look at whether the strands hang down from a variety of thick internal branches like bear's head tooth (*H. americanum*) rather than in one beard-like mass as in lion's mane (*H. erinaceus*).

Take extra time to explore the mushroom's hymenophore. First, identify what sort of structure it has (true gills, false gills, pores, and so on). Then examine it in detail; for example, if it has true gills, see if they are singular structures or whether they split into multiple branches. Pay attention to whether the hymenophore is only on the underside of the cap or whether it extends partway down the stipe. Some mushrooms may also have a small gap separating the interior ends of the gills from the stipe.

Again, a microscope isn't necessary for identifying many common mushrooms. But if you have one, try looking at the spores of the mushroom to see what shape they are; some field guides make note of this, and you can add it to your list of evidence for or against certain species.

Texture

As with plants, start with the visual texture; for example, some species of slippery jacks (*Suillus* spp.) may look wet or slimy even if it hasn't rained recently. The caps of some boletes have a rather velvety appearance. Shaggy manes and scalycaps both get their common names from the flaky appearance of the outside of their caps. The visual texture doesn't always match the tactile texture, so take note of both.

Tactile texture, of course, requires you to handle the mushroom. Some are quite soft and squishy, like jelly fungi or the pores of a *Boletus* that is well past its prime. Others are rather firm; some polypores (particularly Polyporales spp.) harden to a wooden texture, with only the outer, actively growing edge remaining somewhat softer and fleshier.

This is a great time to clear up a common misconception: Unless you are one of the very rare people who are so allergic to some or all fungi that you get external reactions on your skin, it won't hurt you to touch mushrooms. That includes the really nasty poisonous ones. I could pick a deadly galerina (*Galerina marginata*) or a western destroying angel (*Amanita ocreata*) with my bare hands and carry them around all day, and they won't hurt me. The amount of toxins that could theoretically transfer through my skin is so minute that it is negligible. They also wouldn't "rub off" on any edible mushrooms I have in my bag, either, although it's a good idea to keep poisonous or unknown mushrooms in a different vessel anyway just to keep from preparing them with your edibles by accident! The only way those poisonous mushrooms could hurt me is if I ate them and left them in my digestive system long enough for the toxins to enter my liver and kidneys and start causing widespread cell death.

All of which is to say that it's totally fine to handle any mushrooms you find (but it's okay if you want to wash your hands afterward if it'll make you feel better).

Smell and Taste

Most mushrooms are going to have a typically earthy, "mushroomy" smell. However, some may have more distinctive odors. Oyster mushrooms, for example, smell rather like black licorice, while chanterelles have a fruity scent similar to that of apricots. The matsutake (*Tricholoma matsutake*) famously smells like old, dirty gym socks. Give your mystery mushrooms a quick sniff and see if you notice anything distinctive about them.

Some mushroom foragers may take a bite of a mushroom, chew it to get the flavor, and spit it out, perhaps swishing their mouths with water to get rid of every fragment. While this may seem risky, remember that a mushroom must sit in your stomach long enough to be digested for it to hurt you. That few seconds of taste, though, can sometimes offer important information. The peppery milkcap (*Lactifluus piperatus*) is so named because of its sharp peppery flavor; some people dry and powder it to use as a seasoning. However, the sickener's (*Russula emetica*) similar flavor doesn't confer edibility, and this mushroom indeed causes copious vomiting.

A word of warning: A lack of unpleasant flavor does NOT mean that a mushroom is edible! The deadly amanitas, such as death caps or the various destroying angels, reportedly have a rather pleasant flavor when cooked, yet they will still cause massive damage to your liver within a few hours of consumption. Also, while you may taste (and spit out) a small amount of a wild mushroom when it is raw, you need to thoroughly cook your edible mushrooms before eating them. Raw mushrooms can sicken people and in rare cases may be fatal, as in the case of two people who died after eating raw morels at a Montana sushi restaurant in 2023.

Hollow or Solid Inside

While you're cutting your mystery mushroom open to see what colors may be hiding inside, pay attention to whether it's hollow or solid,

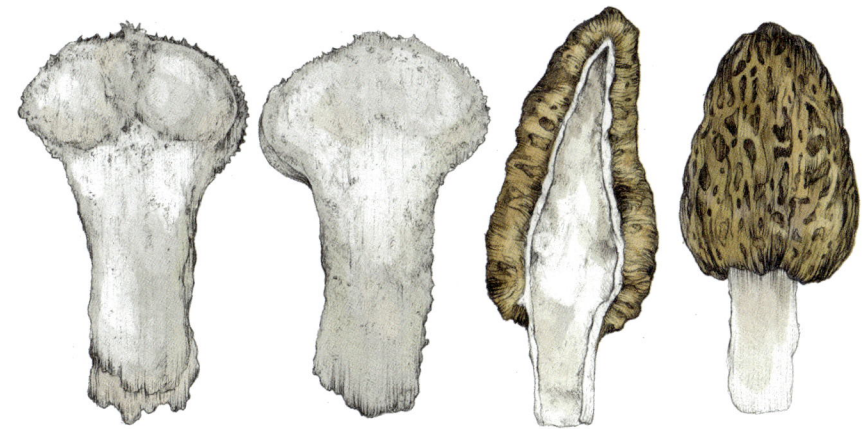

puffball (*Lycoperdon* sp.)   morel (*Morchella* sp.)

in part or in whole. Slice open a true morel and it's hollow all the way through. The false morel (*Gyromitra* spp.) is solid inside, sometimes with some small cavities rather than the single chamber of true morels. The shaggy mane (*Coprinus comatus*) has a solid cap, but its stipe is hollow. The stipes of some larger *Boletus* specimens may acquire some holes in the center, particularly as insects and other invertebrates tunnel their way through the mushroom's flesh.

Sometimes you must look really closely to see if there are any hollow spots or not. Puffballs are solid all the way through. But a young destroying angel or death cap just pushing its way out of the ground looks an awful lot like a small puffball. If you cut one of these highly toxic *Amanita* species in half, you'll notice the gills and stipe just starting to differentiate themselves in a little hollow at the mushroom's base. *Amanita* poisonings may happen because someone mistook them for the edible puffball—all the more reason to examine unidentified mushrooms with care.

Bruising and Deliquescence

Once you cut your mushroom in half to see how solid it is, let it sit for a little while to see it if bruises. Bruising, also known as staining, refers to when a damaged portion of a mushroom changes color. Among the best-known examples are certain species in the Boletaceae, which turn blue when cut, broken, or otherwise damaged. Some turn so dark a blue as to be nearly black! Many blue-staining boletes are toxic, particularly in the genus *Boletus,* although some blue-staining members of the family Boletaceae are safe to eat, so bruising-staining isn't necessarily correlated to edibility. Edible hedgehog mushrooms also may bruise, although they turn orange instead of blue.

Xerocomellus ripariellus bruising blue and shaggy mane (*Coprinus* sp.) showing deliquescence

Wait at least fifteen to thirty minutes after cutting or damaging a mushroom before seeing if it changes color; some change almost immediately, while others need a little more time. There are also some mushrooms, such as the Zeller's bolete (*Xerocomellus zelleri*), that sometimes bruise—and sometimes don't. If you do see bruising on a mushroom, take note, because that will significantly narrow down your options of what it might be.

Deliquescence, on the other hand, is largely limited to mushrooms in the genera *Coprinus* and *Coprinopsis,* such as shaggy mane and common inkcap (*Coprinopsis atramentaria*). They grow like any other mushroom and have a rather firm but flexible tactile texture. But when they reach maturity, the edges of the cap start to roll up and turn black and gooey, causing the gills to spread open and release the spores, until the entire mushroom has turned into a wad of dripping black goo on a stalk. The presence of deliquescence will reduce the number of possible identifications to a handful, so take note if your mushrooms have simply become, well, *mush* in a matter of hours.*

Stage of Development

It may not be immediately obvious to you how old a particular mushroom is. Two species of mushrooms in the same ecosystem may grow at different rates; a slippery jack (*Suillus* spp.) is often full-sized in less than a week, while a northern red belt (*Fomitopsis mounceae*) may grow for several months. Not every tiny mushroom is just sprouting, either; a fully grown fir cone mushroom (*Strobilurus trullisatus*), for example, maxes out at less than two inches tall with a cap less than an inch across.

You'll need to look at other traits to estimate how far along a mushroom is in its developmental cycle. A few things can indicate

---

* If you want to learn even more about how the process of deliquescence works, there's a great article about it at blog.mycology.cornell.edu/2008/07/01 /the-dish-on-deliquescence-in-coprinus-species/.

relative age; used together, they may help you determine whether a mushroom is freshly sprouted or over the hill. We'll cover each of these more generally in a moment, but because the developmental stage can have such a significant effect on the physical traits of a mushroom, I cover it first.

Relative size: While you may not be able to tell whether a given mushroom is fully grown when found in isolation, if you find several individuals of the same species, they may give you a little better idea of its average size range. Note that in mushrooms that cluster together, such as the scalycap (*Pholiota* spp.), all mushrooms in a cluster may have sprouted at about the same time, so for comparative purposes, it is often best to treat a cluster like one large individual and then try to find multiple clusters to compare to one another.

Shape: Some mushrooms may change shape over their developmental cycle. A young oyster mushroom (*Pleurotus* spp.) often looks like a little semicircle sticking out of a decaying log, but as it gets larger, its outer edge can become quite wavy and irregular. A deer mushroom (*Pluteus* spp.) may first emerge looking like a little closed umbrella and then open up into a flat disk when fully mature. Similarly, many umbrella-shaped mushrooms like milk caps (*Lactarius* spp.) have relatively convex or flat caps when young, but as they get older, they seem to turn inside out and become quite concave, like a 100 percent organic, free-range martini glass popping up out of the forest floor.

Color: If you are fortunate enough to find multiple specimens of a particular species, you may notice that their color changes with age. Sometimes it's because of wear and tear; the bright red caps of some *Russula* species fade to various shades of pink after weeks of sun and rain exposure. A mature mushroom ready to produce spores may also show some color shifts; the portobello mushrooms (*Agaricus bisporus*) you find at the store actually started out as white button mushrooms, then became the little brown criminis, and finally large brown portobellos with dark brown gills. Other times color changes are caused by oxidation and decay, like the creamy white pores of a fresh king bolete (*Boletus edulis*) ending up a muddy greenish brown toward the end of its cycle.

Tactile texture: If you've ever bought mushrooms at the store and then forgotten about them until a couple of weeks later, you're likely familiar with the slimy, squishy mess they become when they rot. Since fungi can't poison you unless you eat them, it's okay to handle unknown mushrooms to get a sense of their texture. Most young mushrooms are firm but pliant to the touch and then soften with age until they fall apart and rot. There are exceptions, though; chicken of the woods (*Laetiporus sulphureus*) become harder and brittle with age.

Hymenophore: Many mushrooms protect their hymenophores as they push their way out of the ground or rotting wood. Gills, in particular, tend to be fragile, and so these may not be fully uncovered until the mushroom is approaching maturity. If you look at the underside of a gilled mushroom or cut it open, notice whether the gills are completely exposed or whether part of the cap or veil is still covering them. Some species, such as earthstars or puffballs, have enclosed hymenophores that are never exposed, and the spores are released through an opening in the top of the mushroom that appears only when mature.

Growth Pattern

When you're first looking at a new-to-you mushroom species, pay attention to how closely the mushrooms tend to grow next to one another. Sometimes you will see a huge cluster of them growing out of the same spot in the ground or a rotting log. Other times, they seem to like their space a little more, with one over here, another over there, one way over yonder, and so forth. This is good to note, as you can compare the mushroom's growth pattern to what you see in photos of possible identifications. Keep in mind that a given species may have multiple growth patterns. In the same patch of chanterelles, for example, I may find some specimens widely spaced apart, and then a few feet away a cluster of them all piled almost on top of one another.

Substrate and Habitat Type

So far, we've been looking at the physical characteristics of the mushroom itself. Now let's get into a few peripheral details. The substrate is simply whatever matrix the mycelium producing the mushroom likes to grow in. As mentioned earlier, some species really like soil, while others are found growing on decaying wood, like dying or dead trees. Manure, compost, carrion, and other organic material all host varying species of fungi. Most fungi grow on only one substrate, but this isn't universal. The yellowfoot mushroom (*Craterellus tubaeformis*), for example, will grow on soil or wood, although the wood must be pretty rotten.

Identifying the substrate, then, can be another important clue as to what mushroom you have. True chanterelles grow on soil, as can an inedible look-alike, the false chanterelle (*Hygrophoropsis aurantiaca*). But false chanterelles may also grow on decaying wood, and the poisonous jack-o'-lantern mushrooms (*Omphalotus* spp.), which can also be confused with chanterelles, exclusively do so. Dig around in the substrate a bit; there may be a thin layer of soil overlaying a rotting root or branch, and you want to be sure you can tell exactly which substrate the mushroom is growing out of.

Examine the substrate itself. If the mushroom is growing in soil, what kind is it? Is it sandy, loamy, rocky, clay-heavy, or some combination thereof? Is it damp or dry? Is it anywhere near where animals may be defecating regularly, such as a pasture or barnyard, or near a garden or compost heap? If the mushrooms are growing out of a tree, what kind is it? Identify whether it is a conifer or a deciduous tree; if you can get down to species level, so much the better. It's also a good idea to note whether the tree is alive or dead, as some fungus species can infect still-living trees, especially those with stressed immune systems.

Related to substrate, pay attention to what sort of habitat the mushroom is growing in. Many species prefer forests, especially those that have particular relationships with trees or other plants. However, fungi may grow in just about any habitat type, from wetlands to meadows to deserts. Try to be as detailed as you can. If you found this fungus in a forest, for example, mention whether it is primarily populated by conifer or deciduous trees, as well as which species if possible. The more details, the better, but do what you can with what you know.

Seasons

This is another prime trait for identifying mushrooms. While the mycelium lives and grows in its substrate year-round, it produces its fruiting bodies only at a certain time of year, although some species may have smaller flushes outside that prime time. The oyster mushroom (*Pleurotus ostreatus*) most commonly fruits in fall, but I've visited patches that also produced smaller amounts in spring. Still, if you're trying to identify a mushroom that's popped up in October, and the species you think it might be primarily fruits in spring, you probably want to look at other potential options for identification.

Other factors, such as elevation and drought, can affect the timing of when mushrooms are produced—or whether they appear at all. The fall of 2022 was a rather disappointing mushroom season in the Pacific Northwest because of extensive drought the previous summer that dried out a lot of the mycelium in the area. Since fungi need a certain amount of hydration to move nutrients around for building mushrooms, some of them were simply too dehydrated to produce that year. Other fungal mycelia died off entirely. The following year underwent a cool, wet spring and a relatively milder summer, and the fall mushroom season was more favorable.

Field guides and information websites often list the most common fruiting season for each species of mushroom listed. However, searching for a given species on iNaturalist maps is a great way to get more specific information on when it's typically fruiting in a particular location.

Mycorrhizal Partners

Nature is full of all sorts of mutualistic relationships that fly in the face of the "red in tooth and claw" stereotype. One of my favorites is mycorrhizal partnerships. *Myco* means "fungus," and *rhizo* means "root." When the mycelium of some fungus species grows through the soil, it wraps around the roots of certain plants, particularly trees and woody shrubs. This isn't harmful to the plants, however. Instead, the fungus and the plant share resources. The fungus helps the plant's roots access water and nutrients, while the plant shares some of the sugars it creates through photosynthesis. Plants may also use this network to share nutrients with their offspring and sometimes other plants as well.

Why does this matter to fungus identification when this whole process is happening out of sight underground? Well, in many cases, each mycorrhizal fungus is particular about which species of plant it will partner with. This includes large, highly visible plants like trees. So, if you happen to consistently find a new species of mushroom near or under the same species of tree, there's a chance that they may have a mycorrhizal partnership. Many field guides and other resources will mention any such partnerships a given fungus has. Many of my favorite edible mushrooms, in fact, are mycorrhizal with local trees; I find saffron milkcaps (*Lactarius deliciosus*) and king boletes most frequently amid shore pines, while the Pacific golden chanterelles in my area are particularly fond of western hemlock.

Spore Prints

Unlike the preceding traits, spores take a little more effort to observe properly, but they can be important clues when you're trying to figure out what mushroom you have. Each mushroom has only one color of spore; you're not going to find a mushroom that has rainbow spores! A given species may have a little variation in tone—for example, white to cream or buff to tan. But a spore print should show just one uniform color.

Gilled mushroom and its spore print

To take a spore print, remove the entire stipe of a mature (but not past its prime) mushroom so you have just the cap. Place the cap hymenophore-side down on a piece of white paper; printer paper will do just fine. Put the paper and mushroom in a cool, dry place, and place a mixing bowl upside-down on top of it to keep drafts or breezes from blowing the spores around. Leave it for twenty-four to forty-eight hours. Then carefully lift off the bowl and the cap, and you should have a nice spore print displaying the color quite clearly. There may be a little subjectivity in figuring out the exact color, like the difference between buff and tan, but you can at least rule out mushrooms with green or purple spores in that particular case. If you have pale spores on pale paper and you're having trouble seeing the exact color, wipe some off on black or other dark-colored paper and they should show up nicely.

The spores are not attached to the paper at this point and will easily blow or be wiped off. If you want to preserve the print, carefully spray it with a few layers of hair spray or clear acrylic paint sealer, letting each layer dry thoroughly before adding the next. You can then save it for further identification purposes or personal records, frame it as art, or add it to a scrapbook or art journal.

Pick or Cut?

There's debate about whether it's better to cut mushrooms at ground level with a knife or to pull up the entire thing. Mostly, the argument is whether the mushrooms will grow back faster after a particular method of collection or whether they'll even grow back at all.* When I looked into the matter, the studies and other evidence I found suggested that, for the most part, it doesn't really make much of a difference. One study showed a slightly larger harvest in patches that were harvested exclusively through pulling rather than

---

* This appears to be a by-product of people not understanding the difference between fungi and plants. If you uproot an entire plant, you're likely to kill it. But if you pull up a mushroom, you are not in any danger of pulling up the entire mycelium, which may extend several feet from where the mushroom popped up.

cutting, and the point has also been made that cutting a mushroom and leaving an exposed stump can open the fungus to diseases.

But we're primarily concerned with identification here, and for that reason, I prefer picking over cutting. You want to get as much of the mushroom as possible in hand. Let's say the cup of the mushroom is buried under soil or debris; if you cut off only what's above ground level, you're leaving behind an important piece of the mushroom. Moreover, if you cut off what looks like a young puffball too high up, you might miss the beginning of the gills and stipe of a deadly *Amanita* puffball look-alike.

You can use that shiny new mushroom knife you bought for cutting off the dirty ends of edible mushrooms you've positively identified if you want, but when it comes to identification, pulling is superior to cutting so that you get the whole mushroom.

Notes on Photographing Mushrooms for Identification

If you are going to photograph a mushroom for identification on an app or to share in an online group, it's important to get multiple views of it so that there's more visual information to go on. Here are the basic three views I recommend:

- **Bird's-eye view:** This is when you're looking down over the mushroom, either approaching it or standing over it. It gives you a good look at the top of the cap, and it also allows a good sense of proportion between the mushroom and any surrounding plants or other natural features that can help convey its size (a ruler helps, too). However, this should not be the only picture you take, since many mushrooms look similar from this angle.

- **Bunny's-eye view:** This is a side view, as though you were a rabbit sitting on the ground looking directly at the mushroom. (You can also think of it as the "Smurf

BIRD'S-EYE VIEW

BUNNY'S-EYE VIEW

BUG'S-EYE VIEW

Lactarius deliciosus mushroom

house view.") Here you can see the stipe clearly and how it attaches to the cap. You also get a good look at the proportion of the cap to the stipe, as well as a general idea of the height of the mushroom, again in proportion to other natural features around it. You may want to place a ruler in the photo to help show height.

- **Bug's-eye view:** This may be the most important view of all. It captures the underside of the mushroom's cap, showing whether it has gills, pores, or other structures in the hymenophore. It also reveals what color(s) are found under the cap, which can be another important diagnostic feature of the mushroom; sometimes the difference between one species and another comes down to what's going on underneath the cap.

You are welcome to take more photos than these if you like. Just make sure that all three of these views are included in the photos you're using for your identification.

In the next chapter, we'll meet some close associates of the fungus kingdom, including those engaged in some mutualistic partnerships in which fungi play vital roles, as well as an entirely different kingdom that's often the victim of mistaken identity.

FUNGUS IDENTIFICATION TEMPLATE

Species Observed:
False chanterelle (*Hygrophoropsis aurantiaca*)

What, When, and Where?

DATE: September 23, 2021

LOCATION: Home, Long Beach, WA, shore pine forest east of house

NUMBER OBSERVED: About two dozen

GROWTH PATTERN (alone, in a group): Scattered in small groups

SUBSTRATE (soil, decaying wood, etc.; include species if possible): Growing in well-drained sandy soil

HABITAT: Early-succession shore pine forest with mixed native-nonnative plant community

POSSIBLE MYCORRHIZAL PARTNERS: Shore pines (*Pinus contorta* var. *contorta*)

Other Details

SMELL/TASTE: No distinctive odor noted; did not taste

BRUISING Y/N (if yes, color?): N

DELIQUESCENCE Y/N: N

What's It Look Like?

SIZE: Small, generally <3 inches tall and <2-inch cap diameter

OVERALL SHAPE: Funnel-shaped, although some specimens had flatter caps, especially when younger; Some specimens' caps have a distinctly wavy edge.

HYMENOPHORE TYPE: True gills, which may fork on their way to the cap's edge. Gills are decurrent, growing partway down the stipe.

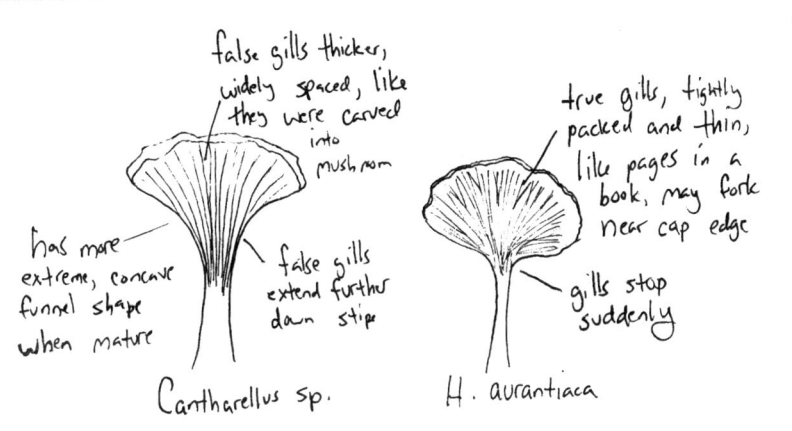

false gills thicker, widely spaced, like they were carved into mushroom

true gills, tightly packed and thin, like pages in a book, may fork near cap edge

has more extreme, concave funnel shape when mature

false gills extend further down stipe

gills stop suddenly

Cantharellus sp.

H. aurantiaca

COLORS/PATTERNS: Upper side of cap, hymenophore, and stipe all a vivid orange. Edge of cap fades to white. When cut in half, interior is pale orange to white.

VISUAL TEXTURE: Soft, maybe slightly rubbery? Gills are crinkly and look a bit like crepe paper.

TACTILE TEXTURE: Firm but flexible; top of cap is smooth; gills are quite fragile and fold or tear with relative ease. Stipe easily snapped.

HOLLOW OR SOLID INSIDE: Solid

SPORE PRINT COLOR: Cream

CASE STUDY

Species: False chanterelle (*Hygrophoropsis aurantiaca*)

DATE: September 23, 2021

LOCATION: Home, Long Beach, WA, shore pine forest east of house

OBSERVATION NOTES: A scattered group of small, bright-orange mushrooms consistently pop up under and near the shore pines east of the house every late summer and fall. They are generally less than 3 inches tall and less than 2 inches across, although in the past I've seen a few larger specimens.

Overall shape is that of a funnel, although smaller ones may have flatter caps or edges that roll under, and some larger specimens have markedly wavy cap edges. The stipe is quite slender. They appear to be rather dry as opposed to slimy and look as though they would be relatively soft to the touch, maybe a bit rubbery. Tactile texture generally matches this assessment; they're easily bent and broken.

The hymenophore consists of true gills, which are thin, fragile, and packed in tightly. Some of them fork toward the edge of the cap. They look somewhat like they are made of crepe paper. The cap, stipe, and gills are bright orange, although the cap fades to white around the edge; when cut open, it displays varying shades of orange. When a spore print was taken, it showed up as a pale cream in color. The mushrooms do not have any particular odor or flavor. No bruising noted, and they do not appear to deliquesce.

These mushrooms tend to grow in small clusters; they aren't growing right on top of one another, but it's common to see a few of them in proximity to one another. Their proximity to shore pines suggests a potential mycorrhizal relationship, given that pines are common mycorrhizal partners. I've primarily found them growing in soil covered in a lot of pine needles, twigs, and other debris.

DISCUSSION: I first observed these little mushrooms back in 2016, shortly after I moved to Long Beach. And I'd identified them as "not *Cantharellus*" for purposes of safe foraging. But it took a few years before I sat down and confirmed my suspicions that they were, in fact, false chanterelles.

False chanterelles and true chanterelles do often fruit at the same time and sometimes in the same habitats; both may be yellow to orange in color and be funnel-shaped. The biggest difference is in the hymenophore. False chanterelles have true gills that look like pages in a book, while true chanterelles have false gills that are thicker, chunkier, shallower, and spaced further apart.* The false gills almost look as though they have been carved out of wax or wood. While both true and false gills may be decurrent, the true chanterelles' gills may travel further down the stipe.

I've also never found true chanterelles over in the shore pine forest, although the false chanterelles love the well-drained soil there. And, as a matter of fact, they are likely to be mycorrhizal with those trees as well. The true chanterelles prefer the older mixed-conifer forest that has a few pines, but plenty of western hemlock.

We don't tend to get many jack-o'-lantern mushrooms (*Omphalotus* spp.) here in the Pacific Northwest, but it's good to rule those out as well, as they can be quite toxic if ingested. Like the false chanterelles, they have true gills, but the gills don't fork and are less likely to have that crepe paper texture that I saw on my specimens. The spore print is yellow as opposed to cream or white. Jack-o'-lanterns are more likely to be found growing directly on dead trees, and they generally won't show up far from their host. Some species are bioluminescent, so if your mushroom is glowing in the dark, it's not going to be a chanterelle, true or false!

---

* We really need to come up with better terms in the natural sciences than "true" and "false" as a way of avoiding confusing sentences like this one.

FUNGUS IDENTIFICATION TEMPLATE

Species Observed:
Orange jelly spot (*Dacrymyces chrysospermus*)

What, When, and Where?

DATE: December 12, 2018

LOCATION: Home, Long Beach, WA, shore pine forest east of house

NUMBER OBSERVED: Three individual clusters or lumps

GROWTH PATTERN (alone, in a group): Clustered close together in a small group

SUBSTRATE (soil, decaying wood, etc.; include species if possible): Decaying shore pine (*Pinus contorta* var. *contorta*) log

HABITAT: Early-succession shore pine forest

POSSIBLE MYCORRHIZAL PARTNERS: N/A; this is a saprophyte that subsists on decaying wood

Other Details

SMELL/TASTE: No distinctive odor or flavor noted

BRUISING Y/N (if yes, color?): N

DELIQUESCENCE Y/N: N

What's It Look Like?

SIZE: 1 inch to 2.5 inches across

OVERALL SHAPE: Lumpy, asymmetrical, short, stubby stipe from which a formless swelling grows; noted faint striations that radiated out from the stipe across the entire cap

HYMENOPHORE TYPE: No gills or pores noted; underside of cap completely smooth

COLORS/PATTERNS: Uniformly bright yellow-orange

VISUAL TEXTURE: Shiny, rubbery, "wet" appearance, but not slimy

TACTILE TEXTURE: Soft, squishy, firmly jelly-like

HOLLOW OR SOLID INSIDE: Solid

SPORE PRINT COLOR: None taken

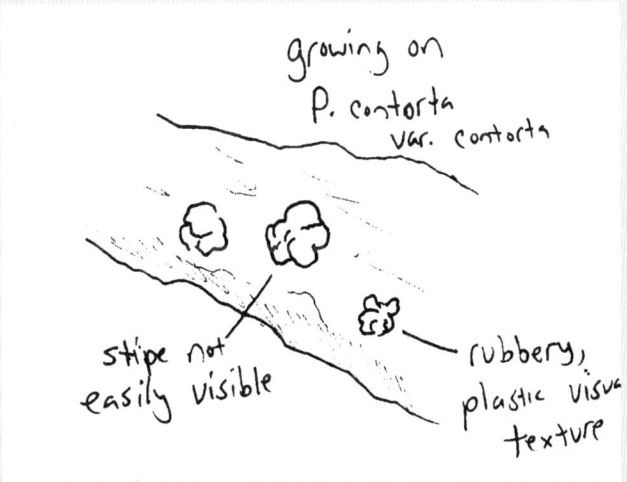

CASE STUDY

Species: Orange jelly spot (*Dacrymyces chrysospermus*)

DATE: December 12, 2018

LOCATION: Home, Long Beach, WA, shore pine forest east of house

OBSERVATION NOTES: These fungi stood out vividly as orange jelly-like lumps on a dead shore-pine log. The largest was maybe 2.5 inches long and less than 1 inch wide. Their color was uniformly a bright yellowish orange that stood in stark contrast to the dark brown wood from which they had emerged. Lacking any real symmetry, each one consisted of a short, stubby stipe and a relatively formless, asymmetrical "cap" that merely seemed to be a swelling on the end of the stipe. The cap had faint striations radiating from the stipe. There was no distinctive hymenophore, as the underside of the cap was as smooth as the top, and they were solid inside with no bruising noted and no sign of deliquescence.

Their visual texture was best described as "rubbery," like a tiny inflatable toy. These little fungi were quite shiny, appearing wet even when conditions were completely dry. They were unsurprisingly soft and squishy to the touch. They did not have a notable odor or flavor.

DISCUSSION: This is one of those cases where my initial identification was incorrect, and in subsequent years, I gathered more information that allowed me to make a more accurate assessment. I looked at these little orange blobs and decided they must be witch's butter (*Tremella mesenterica*), since I had seen it in field guides and other resources before, and it looks quite similar. After I had uploaded it to iNaturalist, another user agreed with my identification, and I thought that was that.

Later on, I learned about the orange jelly spot, which is also sometimes colloquially called "witch's butter" (a good reason to look at scientific as well as common names!). It looks almost identical to *T. mesenterica* in color, shape, and size. The big difference? Orange jelly spot grows only on conifer wood, while true witch's butter prefers the decaying remains of deciduous trees. In reviewing my 2018 observation, I found that the fungi were, in fact, growing on a dead shore pine, so *Dacrymyces* it must be! This is a great illustration of why it's so important to make note of a fungus's substrate; sometimes the difference between two almost identical species comes down to one single detail like this.

CHAPTER 8

Mistaken Identities: Slime Molds, Lichens, and Other Oddities

wolf lichen (*Letharia vulpina*)

WELCOME TO THE NATURE POTPOURRI CHAPTER! This is where I'm going to cover some organisms that either don't fit neatly into the previous three chapters or look like they belong in one taxonomic group when they're actually part of another, which can lead to identification dead ends. I won't be going into quite as much detail as I did on the animal, plant, and fungus kingdoms, but I want to introduce you to them in case you make their acquaintance out in the field.

Slime Molds: Not Actually Fungi

I love slime molds. They're some of the coolest, strangest organisms on the planet, and because they aren't visible for most of their life cycle, it's always a treat when I do get to spot a colony of them. They're most frequently mistaken for fungi because they may look a little like mold and often grow in similar habitats, but they're protists. In fact, *slime mold* is more of a catch-all term for several clades of assorted eukaryotes that don't really fit in anywhere else and don't all share the same evolutionary history.

pretzel slime mold (*Hemitrichia serpula*)

chocolate tube slime mold (*Stemonitis splendens*)

All of the organisms we've looked at in this book are eukaryotes. Eukaryotes include living beings whose cells have a nucleus, and they include animals, plants, and fungi, as well as slime molds and a number of unicellular beings. Prokaryotes, the domains Bacteria and Archaea, make up the remainder of life on Earth, and their simpler cells do not have nuclei or other specialized organelles.

You won't get to see most slime molds in the wild, simply because they're too tiny to be observed without a microscope. The *Myxogastria,* also known as *Myxomycetes,* are a class of slime molds that have an unusual life cycle, including a much larger, more visible stage. They spend most of their lives as single independent cells eating whatever little fragments of food they can find. If conditions become harsh or food is scarce, they can enter a state of dormancy in a form called a microcyst. They can survive in this state for a year, and once things improve, they get back to their microscopic lives.

While slime molds don't have sexes the way we think of them, there are multiple mating types—in fact, one slime mold can have hundreds of mating types. Should two compatible slime mold cells encounter each other, they combine to become a zygote. This is where things get *really weird.*

That zygote doesn't just split into a few new unicellular slime molds. Instead, it becomes a plasmodium, the myxogastric version of The Blob. It devours any organic material in its path and becomes a single gigantic cell that multiplies its nuclei until it's filled with thousands of copies. If conditions are right, the plasmodium stops its all-you-can-eat marathon and becomes a fruiting body. It then produces and releases spores and dies away, while the spores float off to become new unicellular slime molds.

In spite of having no brain or nervous system, plasmodia are eerily intelligent. They excel at finding food, so much so that scientists have experimented with sending them through complex mazes with a treat at the end. Not only do the plasmodia make their way through as quickly as they can ooze along, but they often do so without making any wrong turns.

In the plasmodium and fruiting body stages, slime molds can be easily seen with the naked eye, and the forms different species take are incredibly varied. Here in the Pacific Northwest, you might come across a bright yellow foamy pile of dog vomit slime mold (*Fuligo septica*), little pink spheres of wolf's milk (*Lycogala epidendrum*), or chocolate tube slime (*Stemonitis splendens*), which looks suspiciously like tiny, overcooked corndogs on blackened sticks. I hope someday in my travels that I get to see the pretzel slime mold (*Hemitrichia serpula*), which really does look like one giant crunchy pretzel stretching across a log.

So how do I know when something is a slime mold and not a fungus? It's usually a combination of visual and tactile textures, along with having seen enough fungi to be familiar with the most likely shapes they take. To me, many slime molds tend to appear almost plastic, too slick and perfect to be real. Their tactile textures tend to be squishier than fungi, too; dog vomit slime mold basically feels like a pile of soap foam.

A lot of it comes down to knowing what to look for and recognizing common species even if you haven't seen them in person. Try searching iNaturalist for slime molds in your area to see what people are finding there. I also like to do image searches for slime molds on search engines; while the most common species get the most pictures, sometimes something I haven't seen before pops up.

Parasitic Plants and Horsetails: Nope, We're Not Fungi Either

I briefly mentioned parasitic plants in Chapter 6. These species have figured out how to siphon nutrients from other plants, either directly or by using the mycorrhizal network of plant roots and fungal mycelium. Some are hemiparasitic, meaning that they still acquire some nutrients through photosynthesis, while holoparasites have lost the ability to photosynthesize, and, in fact, no longer develop chloroplasts. Such obligate parasites must have a host plant to survive, while facultative parasites can live without one.

Because many parasitic species are no longer green, people may not always recognize them as plants, especially if they are not flowering. They may appear in a variety of colors and their shapes can be unusual, like fleshy protuberances or cone-like spikes sticking out of the ground; some *Hydnora* species look like black-and-red alien eggs one might find in a sci-fi movie.

Ghost pipe (*Monotropa uniflora*) is among the best-known parasitic plants in North America and is probably the one most likely to be mistaken for a fungus. It grows as a waxy white stalk with a small bell-shaped flower, and it rarely exceeds a few inches in height. Because of its color, texture, and lack of easily recognizable leaves, those first encountering it may assume they've found a sort of mushroom. The same goes for the gnome plant (*Hemitomes congestum*), which appears as small clusters of fleshy pale-pink-and-yellow flowers on the forest floor. I was fortunate enough to be introduced to this unique species when I taught a nature identification class at the Sitka Center for Art and Ecology on the Oregon coast; when I showed it to my students and asked whether they thought it was a plant or a fungus, almost all guessed fungus.

Although parasitic plants may not have all the same structures as other plants, they do at least have a few familiar parts. Look for leaves and flowers, though they may be underdeveloped or strangely shaped. And look for an absence of fungal anatomy, particularly the hymenophore.

ghost pipe (*Monotropa uniflora*)

European mistletoe
(*Viscum album*)

great horsetail (*Equisetum telmateia*)

While we're on the topic of plants sometimes mistaken for fungi, let's talk about horsetails. In some species, their spore-bearing stems emerge before the foliage appears, popping up from the ground like mushrooms. You might be forgiven for assuming the strobili on the ends also indicate these are fungi, and if the stems are brown rather than green, this may further add to the confusion of what kingdom these bizarre beings belong to.

It's safe to touch horsetails, and if you handle them, you'll notice that their stems are much firmer than most mushrooms. Dark rings of tiny, simple leaves called microphylls appear at intervals all along the length of the stem, which is characteristic of these plants. Finally, if you find these odd-looking, club-shaped strobili growing amid clusters of green plants that look like bottle brushes, they're likely all growing from the same underground rhizomes, and so you can be assured that they are flora and not fungi.

Lichens: We Are Fungi—and Then Some!

Unlike the previous two groups of organisms, lichens are categorized within the kingdom Fungi. What sets them apart is the fact that they are composite beings made up of multiple partners, not all of which are fungi. Moreover, lichens are not of a single evolutionary lineage; a lichen is as much a process as it is an organism.

First things first: Who are the partners in Lichen, LLC? The main body of the lichen is a fungus, usually an ascomycete species, and it is called a mycobiont. It creates a framework out of hyphae, like the physical structure of a house. Within that structure lives one or more species of photosynthesizer, also known as the photobiont; most commonly this is an alga, but some lichens have a cyanobacterium instead, and there are even lichens that have both. The photobiont feeds the entire lichen energy made from sunlight, and the mycobiont provides water and environmental nutrients.

For a long time, scientists thought these were the only partners found in a lichen. This led to naturalists like me using corny jokes to help people remember them, like "When a fun guy meets

Top to bottom: dog lichen (*Peltigera canina*);
pixie cup lichen (*Cladonia* sp.)

an al-gal, they take a *lichen* to each other!" And then—it became a love triangle. A few years ago, studies found that a basidiomycete yeast was a hidden third component of some lichen species. Its role is not fully understood; it may produce certain chemicals such as vulpinic acid, or it might be the architect who determines the lichen's shape. The discovery of the yeast has also led some scientists to speculate that a lichen may be a holobiont, a miniature ecosystem that revolves around the structural fungus.

Many species of fungi, algae, and cyanobacteria are found participating in lichens; some species also live as independent entities. Each partner in a lichen contributes something useful to the whole community, and everyone benefits. Lichens can survive in harsher conditions than the individual organisms can tolerate. *Cladonia rangiferina,* for example, is known as the reindeer lichen because during cold Arctic winters it may be one of the only foods available to caribou (*Rangifer tarandus*).

Sometimes environmental conditions change to such a degree that if one of the partners can be more successful independently, it will split off and go its own way. Some lichen-forming fungi are so dependent on the collective to fulfill certain biological functions that they have lost the genes to perform those tasks themselves. Certain lineages of independent fungi and photosynthesizers may have had to join up and split apart multiple times over millions of years. If a species that previously lichenized spends too much time alone, though, it may lose the ability to lichenize entirely.

Because lichenization is a complicated process with a multitude of participants, it's easiest to just categorize lichens as fungi because those are the most recognizable partners. How do you differentiate between a lichen and any other fungus? Texture is a big giveaway; most lichens are rather dry in appearance and often feel crusty, leathery, or brittle. Lichens also don't have a big color range compared to fungi. A pale bluish-green or gray is the most common shade seen, and yellow and orange are also common. A few, like wolf lichen (*Letharia vulpina*), are a bright acid green, while some species of *Peltigera* and *Lobaria* may display somewhat more subdued shades of green and gray. Beyond that, lichens lack the sheer diversity of green hues of the plant world or the wider rainbow of shades seen in mushrooms.

They are also not especially large. Most individuals are only a few inches across, although there are a few exceptions, with the world's longest—up to twenty feet long—being *Usnea longissima*. Generally, lichens don't grow quickly, sometimes only a fraction of a millimeter a year; because of this, I recommend not taking lichen samples unless they have already fallen off their substrate on their own.

Speaking of substrate, lichens will grow on just about anything if they hold still long enough. Some prefer stone, while others colonize the bark of live trees or the remains of dead ones. Lichens can be found growing on soil as well and are an important part of biocrusts on the surface of soils in arid areas. If you don't wash your car for a while and live in a sufficiently humid climate, you may end up with lichens colonizing convenient crevices (my old Toyota Corolla has a nice little *Cladonia* colony growing on the frames of the side mirrors).

Most tellingly, lichens tend to grow in one of a few particular patterns. Here are the most common ones you're likely to encounter:

- **Fruticose:** Has a thick, bushy appearance; the individual "branches" are broader and more dimensional than those of filamentous lichens.

- **Squamulose:** Appears to be made of multiple overlapping layers, like miniscule shingles or scales.

- **Crustose:** As the name suggests, grows like a crust over whatever substrate it prefers.

- **Foliose:** Leaf-like lichens that may look similar to squamulose, but the individual "leaves" are usually larger.

- **Filamentous:** Consists of several slender root-like filaments, with smaller ones branching off them.

There are other growth forms, of course, and some lichens even display multiple forms at the same time. Lichens reproduce asexually using isidia, which are tiny structures a millimeter or less in height that can break off to become new lichens, and soredia, microscopic masses of mycobiont hyphae and photobiont cells that may appear en masse as a powdery texture on the lichen's surface; a cluster of soredia is known as a soralia. Sexual reproduction in lichens may be achieved with any number of specialized structures such as the disc-like apothecia and flask-shaped perithecia. You may need your magnifying glass to see these various reproductive structures in detail, but their distinct shapes may be helpful in lichen identification.

A few lichen species produce mushrooms like those of an independent fungus. The first one I ever encountered was a lichen agaric (*Lichenomphalia umbellifera*. At first glance, it appeared to be just a tiny, pale umbrella-shaped mushroom growing out of a patch of moss on a rock. The "moss," in fact, was a rather unassuming little crustose lichen, whose dark green color came from its algae photobiont. It was a fascinating specimen to explore, and it makes me wonder if it represents a transitional phase in the process of lichenization where the mycobiont is still independent enough to produce its own fruiting bodies rather than those more commonly seen in lichens.

One final thing about lichen identification: It's often quite difficult to identify a lichen down to the species level, especially microlichens. Many of them look quite similar, and differentiation among species may come down to careful examination of spores and other microscopic features or how the lichen reacts to contact with certain chemicals, such as potassium hydroxide and calcium hypochlorite. I've never taken my lichen identification to that level, and I primarily stick to more easily accessible traits, with the acceptance that sometimes the best I can do is come up with a probable genus or two.

Microscopic Beings

I'm going to preface this section by saying that I don't have a lot of experience in peering at tiny organisms through microscopes, so I won't spend too much time on this topic. Still, there is a subsection of everyday naturalists for whom the tiniest beings hold the biggest wonders. Depending on how powerful your microscope is, you might get to see micro-animals like tardigrades, rotifers, and nematodes, or single-celled algae floating in a drop of water. Even more powerful lenses may get you a glance at some bacteria or archaea. However, identification may be more challenging than with larger organisms.

A good starting point if you're new to amateur microscopy is MicrobeHunter Microscopy (MicrobeHunter.com). This site features videos, a podcast, a forum where you can ask questions and chat with other micronaturalists, and a shop link to an array of Amazon listings, including recommended books and suggested microscopes and supplies. It's a great starting point if you want to get involved with examining and identifying microscopic organisms.

One book that I didn't see on the aforementioned list that I think would be a great additional resource is Betsey Dexter Dyer's *A Field Guide to Bacteria*. It features more than three hundred pages full of information about how to observe bacteria in everyday settings at home and outdoors. I especially like the sections on the field marks that can be used to help identify various bacterial taxa, including visual and aromatic traits and preferred habitats. It's the only field guide I've found that focuses on such small beings, and it doesn't require you to have a background in microbiology.

Dyers goes into detail not just about how to see individual bacteria under a microscope, but also where you can observe entire colonies with the naked eye. Think about the last time you found mold on a slice of bread or other food or growing on a rotting log; you were seeing significant colonies of unicellular fungi. Powdery mildew and other fungal plant diseases show up as spots or coatings on leaves, twigs, and other parts. The massive, knotted burls seen on some trees are evidence that it has survived an infection from bacteria, viruses, or other pathogens. And if you've ever seen a tangle of branches growing from one spot on a tree in a "witch's broom" configuration, that is often caused by phytoplasmic bacteria introduced by insect vectors.

Not every collective of microscopic beings is engaged in parasitism, though. The star jelly (*Nostoc commune*) isn't a jellyfish or other animal, but a cyanobacteria that forms large colonies that can be several inches across and may be mistaken for seaweed. Algae blooms occur when there are extra nutrients in the water, and microscopic algae populations explode as a result. If you see discolored water in a hot spring or around a stagnant pond, that may indicate a colony of bacteria or archaea. Some of these display bright colors, such as the vibrant purple of bacteria in the order Chromatiales or yellowish *Sulfolobus* archaea that thrives in the acidic hot springs of Yellowstone National Park.

Not all microscopic aquatic beings are extremophiles, of course. Any freshwater pond or stream is likely teeming with life at the smallest scale, and the ocean, of course, is full of numerous species of plankton. Soil is full of bacteria, archaea, and other microlife as well, and the *Do-It-Yourself Guide for Microscopy of Agricultural Soil* by Katelyn Solbakk of Mikroliv is a good guide to extracting them from the dirt; it can be read in full at her website, Mikroliv.no.

So, when you're out exploring in nature, chances are you're viewing vast communities of microscopic beings—or the remnant of where they once lived. If you have a microscope at home, you might scrape off a tiny bit of a discolored leaf or put a small sample of water or soil in a bottle, and then see what you discover under the lens.

CHAPTER 9

Troubleshooting: Confusing Critters, Wrong IDs, and More

inland wood oats (*Chasmanthium latifolium*),
eastern deer mouse (*Peromyscus maniculatus*)

WOULDN'T IT BE GREAT IF YOU WERE ABLE TO EASILY identify every living being you came across with a minimum of fuss? Unfortunately, that's not going to happen for anyone, not even the most experienced field naturalists with decades of experience. While you may be able to positively identify some species the first time without much effort, you're also likely to find yourself stumped now and then (and not just when identifying trees!).

This chapter is about some of the common challenges in nature identification and potential workarounds for them. It doesn't mean you aren't going to still run into difficulties, but hopefully I'll be able to address at least some of the likely problems.

I Didn't Get a Good Look at It and My Photos Were Terrible or Nonexistent!

Let's say you're observing an animal that just will not stay still. Maybe it's a bird flitting among the branches of the trees overhead, a fish that insists on hiding under a log, or a particularly skittish butterfly that flies away every time you get within ten feet of it.

If you can't keep your eye on it long enough to get a good sense of its field marks and other traits, you likely won't be able to get any good pictures, either. And if you're stuck in a situation where you don't have the time to try to follow the creature, you're basically limited to that initial glance. So how can you compensate for a lack of information?

First, learn to make the most of what you do have, and *write it down*. I am a big advocate for recording my observations of an unknown species as soon as possible. Unless I am driving or otherwise occupied, I immediately note whatever colors, sizes, shapes, and other traits I managed to catch, even if they're estimates or general descriptions. It can be tempting to second-guess yourself as you start doing research. Trust your initial impressions; the human memory is quite malleable, and you can easily fool yourself into thinking, "Ah, the orange on that butterfly wasn't *that* bright," or "I thought maybe the lizard's body was lighter around the edges." This is a big part of why it is so important to be detailed in your initial description and to record it as soon after the sighting as you can.

However, there are likely to be times when you're going to be working with a split-second observation that allowed you to recognize only a few general features of the animal. Again, I emphasize recording everything you did notice as soon as you can after the observation, because that's all you're going to have to go on once you start researching potential species.

What about trying to get a better look? If you have the opportunity to follow the animal without harming it, by all means go for it! However, I caution you to be careful in assuming what "harm" may entail. Even if you don't ever physically touch the animal, it may become quite stressed by what it perceives as you pursuing it. Some animals may do best if you simply sit still for a bit and allow them to come back into view; this involves patience, but it can reward you with a chance to observe the animal's natural behavior for a longer period.

This isn't the only sort of situation where you may not have enough time to thoroughly observe your specimen. Plants and fungi may stay in one spot, but if you're in a hurry somewhere or if you're on someone else's timetable, the best you may get is a quick glimpse as you pass by. Go back on your own time if you're able to and see if you can find what you were looking at before. The sooner you can return, the better, especially with fungi, as some mushrooms will persist for less than a week before decaying into a pile of rotting goo.

Sometimes you aren't going to get the quality of observation you really want. The best thing you can do in that case is make peace with the reality of the situation and hope that in the future you'll have another chance to get to know that mystery species in more detail.

It's Not in Any of My Field Guides!

As much as I love my ever-growing collection of field guides, they are necessarily limited in scope. There's no guide to every single organism of the world, and most field guides focus on a particular group of taxa within a specific region. Many field guides include the most common species and may not cover more obscure ones that are seldom seen in the area. Field guides primarily focus on native wild species, which means if you're looking at a number of domesticated or otherwise nonnative species in your area, you're not likely to find them in your books.

Field guides also have limited space for photos or other images of a given living being, especially if they're trying to cover as many species as possible. A quick online search, on the other hand, can turn up thousands of pictures, including several that look exactly like that sparrow you observed in real life. This holds true not just for birds and other animals, but also plants, fungi, and so forth.

I cannot overemphasize this enough: Use as many resources as you possibly can to gather the maximum amount of information on the species you think you may have found. If one field guide doesn't have the organism you saw, try another until you've exhausted the books you have access to. After that, it's time to move on to other tools. If you have a photo of what you saw, upload it to an identification app to get some suggested species to start with. Search for websites that may describe the species you're trying to identify. Type a basic description of it into a search engine, and see what websites and images pop up as a result. Post any pictures and detailed descriptions you have to relevant groups online, or send them to experts at universities and other such entities.

I Can't Decide Which Species It Is!

This is a pretty common problem, actually. Remember what I was saying earlier in the book about little brown jobs, little brown mushrooms, and damned yellow composites? Those groups exist because *everybody* has trouble with them, not just novices. Just about every group of organisms is going to have multiple species that are incredibly similar, and in some cases differentiating between the two may involve dissection, microscopes, chemical compounds, and other detail-oriented tools.

Just as a medical professional must go through the process of differential diagnosis when a patient has symptoms that match several potential conditions, the everyday naturalist needs to be able to compare similar species and determine which one is most likely to be the species observed. That's exactly why I went into so much detail in the previous chapters about traits like colors, behaviors, habitats, and so forth. It's not uncommon for the deciding factor to be something seemingly insignificant, so you need to compare all the information you gathered on a mystery organism with all the potential species it could be.

Let's say I have taken a trip down to the very southwestern corner of New Mexico, and I see a bright-yellow fuzzy caterpillar with several longer tufts of black hair on its back and a black head and feet poking out from all the fluff. In looking up potential species, I narrow my options down to either the American dagger moth (*Acronicta americana*) or the cottonwood dagger moth (*Acronicta lepusculina*). The American dagger is much more widespread and common and has been found in the region of New Mexico where I made my observation. The cottonwood dagger, on the other hand, seems to rarely be observed in the southernmost quarter of the contiguous United States, nowhere near where I had been. While it is still possible that a few cottonwood daggers may exist in southwestern New Mexico, it is much more likely that I found an American dagger caterpillar.

If you're having trouble choosing between species, go back to the relevant chapter in this book and use the sections there as a checklist of things to compare—color, size, shape, and so on. Review the entries in field guides and other resources on each of the potential species you're looking up and ask other naturalists for their perspectives as well. As I mentioned earlier, if you're getting stuck trying to tell Species A from Species B, there's a good chance other people are, too. Remember that many species have different appearances according to sex, age, or regional population. It is also possible that none of the species you currently think you've found are the one you observed, and the answer lies with yet another contender with more variation in its ranks. If none of the options you've chosen so far are a solid fit, set them aside and research other possibilities.

I'm Someplace I've Never Been to Before, and I Don't Recognize Anything!

I know many people who have found themselves in this situation when traveling or moving to a new place. If you're at least on the same continent you're familiar with, you may find some species you know, but if you're now halfway around the world, it's likely that you're surrounded by a totally new cast of ecological characters.

If you know you're heading into the unfamiliar, do some research on local species ahead of time; you can pick up a few field guides relevant to the area and bookmark websites on its natural history so you can access them later in the field. There may even be online groups on Facebook and elsewhere dedicated to the flora, fauna, or fungi of your destination where you can ask for suggestions. Set aside time before you leave for your trip to peruse these materials; you don't need to read every field guide cover to cover, for example, but it's good to at least page through and look at the pictures. That way, if you see something out in the field that you recognize, you'll know you can go back to the books and other resources to find out what it was.

What happens if the guides are in languages you don't understand? You can copy sections of ebooks and websites and paste them into Google Translate. The Google Play Books app also has a built-in translation tool; highlight the text you'd like translated, tap Translate, and the app will give you the option to choose the "to" and "from" languages. Do be aware that AI-based translation tools are imperfect, so verify everything with other sources whenever possible.

If you have the money for it, consider hiring a local guide. Ecotourism is a growing industry, and it's easier than ever to find a reliable expert to show you around and tell you what organisms you're seeing. Online groups, search engines, and travel agencies may lead to multiple potential options. Airbnb, Expedia, Tripadvisor, and other travel-related companies now allow you to search for activities and experiences, and GetYourGuide.com is solely dedicated to connecting people to professional guides.

Once you're at your destination, take as many pictures and notes as you can. You may not be able to identify everything in the field, but you can at least learn about them later. Perhaps you can do a little online research in the evenings or while traveling to your next destination.

Compared to travel, moving to a new location gives you the luxury of time to identify and learn more about the new and unusual animals, plants, and fungi you encounter. You may well find that you already know some species, either because they were also found where you moved from or because you've seen them in books, online, and so on. It's still a good idea to double-check your identification because there can be look-alikes. For example, if I moved from the Pacific Northwest down to Florida, I might be lucky enough to encounter a Florida scrub jay (*Aphelocoma coerulescens*). If I just assumed it was part of a particularly easterly population of the California scrub jay, which looks almost identical, I would be incorrect.

Just as you need to make new friends and other connections, so you'll also have to take time to be introduced to your local "nature neighbors." Start from square one: Gather resources, make time to go out and observe living beings, and then go through the process of identifying each one. You may get lucky and find a familiar face (or frond or fungi), but let yourself be excited by the prospect of being able to positively identify even more species than before!

There Aren't Any Resources Specifically for My Area!

I am privileged here in the United States to have access to an embarrassment of riches when it comes to both traditional and digital identification resources. I don't have an exact number of the field guides focusing on our flora, fauna, or fungi that have been published over the years, but I'd be willing to bet it's in the tens of thousands. Not only are there general guides that cover North America, but my collection also includes titles that cover certain regions of the United States, individual states, and even specific metropolitan areas, areas of public land, or single ecosystems. While few guides on unusual beings like lichens and slime molds have been published, you could probably fill an entire bookshelf each with field guides on American birds, native plants, or mushrooms. And that's not even getting into the many websites, ID apps, and other resources also available.

However, I am mindful of the fact that many regions of the world don't have this wealth of printed information. If that's the case for you, start by seeing what resources you can find. Look for any field guides or other books on local natural history; sometimes a search on Amazon will bring up titles, although not all of them may be in print or available secondhand. Check the bibliographies of any books you can access, as more possibilities may be listed there. Search online for relevant websites; you might search for animals, plants, or fungi of your country or of the nearest national park or wildlife preserve if there is one.

Libraries are another good option if you have one nearby, or if you can call or email the closest one. Even if they don't have a particular book available in their collection, they might be able to suggest other titles you can look for online. The same goes for museums, colleges and universities, zoos, and any other institution involved in researching or promoting the natural history of your region. Ask around your community to see if there's anyone who has a lot of knowledge about animals, plants, and so forth. Even if they aren't formal educators or other professionals, sometimes you may find other everyday naturalists who would love to have someone else to share their love of nature with.

Search the internet for any websites about your local ecology; even if they don't cover identification of individual species, they likely still contain valuable knowledge. If you contact a website's creator, they may be able to suggest more resources to explore. Contact any scientists or authors who have written about the natural history of your area, even if they live in another country; see if they have written any books or papers, or if they can refer you to other materials that can help you with identification. If they don't speak your language, don't be afraid to use a translation app or website to aid in communication.

Online groups like those on Facebook are another option. Even if there aren't any tailored to your area, go to more general groups, such as those on plant or bird identification on a particular continent or even worldwide, and see if anyone else is in your area. Try creating a group that's specific to your region; if there currently aren't any, try to keep yours general, such as "Natural History of [Your Country]" rather than "Migratory Shorebirds of [Your Country]." With luck, others will join, and you'll have created a good local resource; if the group gets big enough, you might consider making a few more groups that cover regional taxa that members are especially interested in, like birds or mushrooms.

Unfortunately, some areas just may not have many resources, and not everyone will have the energy and time to create an online group or to contact experts. Just do the best you can with what you have and keep an eye out for news about any books, websites, or other materials that may help you with identifying local species.

I Don't Have Anywhere Safe to Go!

Some locations are heavily developed, with few transportation options that go to greener places. While urban species certainly offer opportunities for observation and identification, the biodiversity just isn't going to be as great as in a wilder location, and it may be that you've simply run out of new species to explore. If you don't drive, see if you can catch a ride with someone you know or take public transit to a park or other green space. Some organizations of nature enthusiasts, like local birding groups, may also offer carpooling to members who lack their own transportation.

It's also possible you live someplace where it isn't safe to venture outside because of the threat of violence or other serious hazards, and you don't have the option to leave even temporarily. Your safety needs to come first, and as enjoyable as nature identification can be, it's not worth losing your life over. That said, many outdoorsy people in war-torn countries, dangerous neighborhoods, and other hazardous settings have had to weigh the risks and decide that it was worth finding ways to enjoy nature, even if it wasn't ideal. As someone who is admittedly in a quiet place and who takes her safety for granted, I can't make that decision for you, but I wish you all the great discoveries and none of the tragedies.

Finally, some minorities face barriers in accessing the outdoors due to biases and serious safety concerns. While I have no compunctions about heading out onto a trail by myself, I know many women who either won't go alone or don't even venture further than a public park because of the risk of attack. Racial minorities have been frequently marginalized in outdoor spaces, and most people hitting the trails are white.

Some organizations are working to raise awareness about these imbalances and promote opportunities for change. Women Who Hike features hiking groups across the United States, as well as Canada and Europe; Trail Dames encourages "women of a curvy nature" to join group hikes at a leisurely pace, allowing for plenty of enjoyment of the surrounding nature. Outdoor Afro, Outdoor Asian, and Minority Outdoor Alliance are just a few of the growing number of initiatives to promote racial diversity outdoors and offer safe recreation experiences. More parks and other public lands are also reaching out to minority communities to make their spaces more welcoming.

Everyone deserves the opportunity to safely spend time outside, whether for nature identification or not.

I Don't Have a Lot of Money for Resources!

No problem! You don't have to have a big bookshelf full of field guides or $2,000 binoculars. This is another one of those "work with what you have" cases. If you have an internet connection, you already have access to many great websites and groups; even an older smartphone can use many of the free ID apps out there like the Audubon Bird Guide, iNaturalist, or Merlin.

If you have a local library, see what they carry in their collection or through interlibrary loan; some libraries even have backpacks full of binoculars, field guides, and other tools that you can check out. Used bookstores and thrift shops are other options for resources on a budget, and ebooks are often cheaper than brand-new paperbacks or hardbacks. For those who can access local birding and other naturalist groups, see if anyone is interested in holding an in-person books-and-binoculars swap where people bring things they no longer need, pick out stuff they can use, and donate any leftovers to charity.

Finally, if you're connected to other people online or in person, ask them if they can give you information from books or other resources they have or otherwise help you confirm your identification. And, of course, be willing to return the favor if someone asks; you just might make the difference in someone else's ability to figure out what cool thing they've seen!

I Think I Got the Wrong Identification— What Do I Do?

Because a significant subsection of my readership is likely to be foragers because of the classes I teach, let me remind you: If you are not absolutely sure that the plant or mushroom you have is an edible species, DO NOT EAT IT. If you've eaten something and only afterward realized that it wasn't what you thought it was, try to figure out what you ate, watch for symptoms like gastrointestinal upset, and err on the side of getting medical attention.

Now, in a more general sense, if you got your identification wrong, it's okay! Sometimes, the right answer may be presented to you by someone you're talking to or a new piece of information you've received. Otherwise, just start over from the beginning as though this were a new organism you've never tried to identify, and check your work as you go along. Here are some things to consider:

How thorough was your observation? Did you miss any details the first time through that might help you get a better identification? If you've taken photos or otherwise recorded details, go back over them and see if there's anything you've overlooked. Focus especially on any traits that may be likely to cause confusion. New World sparrows, for example, are best identified by the markings on their faces, and a species identification may come down to the presence or absence of one particular line of color on a bird's cheek. Take your time and don't make assumptions based on your initial assessment; pretend this is your first look at this specimen.

Did you use every resource you had access to at the time, and have any new ones become available since your initial identification? Look for recently published field guides, new websites or additions to existing ones, and naturalists or other experts you haven't contacted before. Remember that species can be reclassified into a different genus or family, or even split into multiple distinct species, so make sure you're researching both the old and new taxonomic classifications.

How consistent and reliable was the information you got? Is there a chance there may have been a mistake in one of your sources, or that new information may make an older source obsolete? Generally speaking, newer references will be more reliable than old ones, and professionally edited material like field guides and other books are more trustworthy than casual websites or social media, but gauge each source on an individual basis. After all, books can have errors, and sometimes that person who replied to you on Facebook turns out to be a solid authority on the specimen you're asking about.

Did you have any trouble using available resources?
Something like a dichotomous key can be challenging for many people to use because they don't fully understand the technical terminology used or how to identify precise anatomical features of an organism. Either try to get some help in understanding how to best use a given source, or set it aside and focus on those you can use with more confidence.

If you end up getting an identification wrong, please don't be hard on yourself. Naturalists of all sorts make errors now and then. Charles Darwin himself posited in his most famous book, *On the Origin of Species,* that bumblebees (*Bombus* spp.) were the only pollinators of red clover (*Trifolium pratense*). He found out after publication that honey bees (*Apis* spp.) also pollinated clover. He subsequently wrote to a friend, "I hate myself, I hate clover, and I hate bees." I think we've all been there, Mr. Darwin.

This is one of those times when I am especially prone to asking other people for feedback. I'll contact other naturalists or post on a group, with a photo and detailed information about the organism I'm trying to identify. I also mention what I thought it was and why that turned out to be wrong. With luck, someone else will be able to suggest a more accurate ID, which I can then compare to what I saw to make sure it's a good match.

I Don't Feel Confident Enough in My Identification Skills!

This is a pretty common experience, so you're definitely not alone. Nature identification can be quite overwhelming because of how many tools are available and how many skills you need to hone—plus, of course, all the many and varied species out there to be identified! Add in that many of us tend to put pressure on ourselves to not make mistakes, and that equals plenty of people who are concerned about getting the wrong answer.

Here's the thing, though: I'm not sitting here grading you and there's no minimum standard you need to meet. While some of you may be using this book to brush up on some of your identification chops for work or volunteering, and others may have been assigned it as part of a class, you shouldn't be expecting perfection from yourself. The truth about nature identification is that there are always going to be errors. And we often learn more from our mistakes than our successes; they show us where we need to put in a little more effort and focus so we can become better.

I also want you to get comfortable with the phrase "I don't know." I say it a lot! Including in my classes, where I'm supposed to be *the* authority in the room (which, by the way, is kind of silly, because not only am I just one of several everyday naturalists there, but I also always learn some neat things from my students even as they're learning cool stuff from me). "I don't know" offers you the ability to stick a pin in something and save it for later. It means you can wait until you have more time to follow up on something you're not sure of. And it never has to be permanent; there's always the possibility that an "I don't know" will someday be replaced with an "Aha! I've got it!" Finally, it prevents misinformation because it tells others there's some uncertainty, rather than boldly promoting an inaccuracy as truth. In short, a little caution isn't going to hurt you.

That said, you also shouldn't veer in the other direction and always doubt yourself. I have taught many people various sorts of nature identification over the years, and few of them had any professional experience. Most of my students have been the sort of everyday naturalist with little to no formal training, or maybe some casual practice, but in some cases no experience at all. It really is a skill that I feel most people can access, and it's a matter of finding the methods and tools that work best for each person.

Here are a few things that may help build up your confidence:

Spend time getting better at noticing things in nature. This is something I've done almost my entire life, and it's absolutely a skill that can be cultivated. As an increasingly indoor-oriented species, many of us only really go outside when we're traveling from one constructed space to another—our home to our car, our car to our workplace, and so on. Because we tend to tunnel-vision on getting to a destination (and we may be busy paying attention to the road!), we often don't notice what's going on around us in the world of nonhuman nature. So, try taking a walk or sitting outdoors with your only intent to be in that place and seeing how many individual species you notice. If you want, you can count how many you find on a given block in your neighborhood, on a specific stretch of trail, or from wherever you happen to be sitting today. You don't need to know their names, only that this one is different from that one. Just by making these basic observations, you are already training yourself to pay attention to the traits that make each species unique.

Sometimes it's easier to fix someone else's problems than work on your own. If you're on any identification groups online, scroll through posts where people have asked for help with identification, and see if you can figure out what they've found without looking at the comments. You don't have to put down your answer if you don't feel comfortable; this is just an exercise for your practice. Once you have your identification (or an "I don't know"), check the comments and see what other people think, then see if you agree with them.

Practice with something that's relatively easy, like a species you already know or that is commonly found in the area. Here in North America, I might recommend the American robin, the common dandelion, or the portobello mushroom (the latter with which you can make a nice spore print). Sure, you already know the answer to the question, but you're practicing doing the work that would be necessary to arrive at that answer if this were an unfamiliar being. Try to resist the urge to go, "Oh, I know that the robin is a thrush!" or "This is definitely an edible mushroom because I bought it at the grocery store." Act as though you know literally nothing about this organism except what you observe, and then start using your skills and tools to make your way to a positive ID.

Create a report on a known species, as though you were trying to explain to someone else how they would recognize it if they came across it in the field. You can use the templates I've included in the book as an outline if you like. Feel free to produce the report in whatever format you prefer, whether that's filling out the template, writing an essay, illustrating in a sketchbook, or even making an audio or video recording. One thing that really helps is pointing out any similar species that you're aware of and noting how to tell the difference. If you want to see some examples I've done, check out the How to Identify series of articles I've written on my website at RebeccaLexa.com.

Finally, be patient with yourself! It's a marathon, not a sprint, and you're the only person you're competing with. In fact, don't even think of it as a competition, but instead as an adventure. It's your opportunity to learn more about the amazing biodiversity of Earth and all the incredible beings we share this little green planet with. You don't have to learn a certain number of organisms in a day, a week, or a year. You get to do this at your own pace, fueled by your curiosity and not some external pressure.

And chances are your confidence will grow as you get more practice under your belt. The more you practice nature identification, the more it will become automatic for you to notice the various traits and other details about what you're observing. It's like learning to drive a car. When you first get started, you must consciously think about which pedal is which, what each of the little doohickeys on the steering column do, when to check your mirrors, the various rules of the road, and so forth. But after you get used to it, you don't have to think about the bulk of all that—you just do it! Sure, you still need to be consciously mindful of the road and other drivers, but you no longer need to think, "Okay, the pedal on the right is the gas."

Some people find it helpful to give themselves more reasons to get out and learn about their local nature. The next chapter covers one of my favorite excuses to go outside—citizen science!

CHAPTER 10

Using Your Powers for the Forces of Good: Citizen Science and Science Communication

fiddler beetle (*Eupoecila australasiae*)

IF I ASK YOU TO PICTURE A SCIENTIST AT WORK, what's the first image that pops into your mind? For many people, it resembles those stock images of a person in a clean white lab coat looking at a vial of some colorful liquid in a test tube or peering into a microscope. There are certainly plenty of scientists working in labs in a variety of fields (and I can tell you from personal experience it's fun to listen to them point out all the errors in those staged photos!). However, you could just as easily imagine a scientist sitting behind a desk at a computer or reading through books and papers.

And then there are field scientists who focus on biology, ecology, and related topics. While they may spend some time in a laboratory setting or office, they also dedicate as many hours each year as they can to studying living beings in their habitats. This might involve surveying a given species within a designated area, observing animal behavior, or looking for rare or undiscovered species. Some focus on the habitats themselves, examining what qualities make each one unique, determining how much damage has been done by human activity, and even experimenting with ways to return a place to its natural balance.

The field scientists in charge of a particular project generally don't work alone. They may have colleagues or students who join them on-site and help with collecting data. Sometimes they bring in citizen scientists to assist with tasks that don't require years of formal training but who are still crucial to successful fieldwork.

A citizen scientist is a member of the general public who participates in a scientific endeavor, often by gathering data that professional scientists can then analyze. Their contributions mean a greater amount of information can be collected; in fact, some projects are primarily driven by the input of citizen scientists.

Citizen science is not a recent phenomenon. Many participants in Western science over the past few centuries were trained in nonscientific fields but followed their curiosity about the world on the side, like cloth dealer Antonie von Leeuwenhoek, who developed his own lenses for assessing the quality of his wares and, in the process, revolutionized the field of microscopy. Some early Western naturalists were wealthy nobles or merchants who collected natural and cultural artifacts brought back from around the world and then set about trying to understand and categorize them; their cabinets of curiosities were the predecessors of today's museums.

Today, one need not be backed by intergenerational wealth or well-off patrons to participate in citizen science. Anyone with a little extra time may be able to volunteer for a growing number of projects. While some of these occur in person at a specific site, others require only a smartphone with an app for recording data. There are even projects that need volunteers to check in on live video feeds of animal dens or other sites to monitor any activity, something that can be done from the comfort of one's home.

Citizen Science as Science Communication

The older I get, the more firmly I believe that scientific literacy is a crucial skill for everyone. This was confirmed for me during the COVID-19 pandemic, which provoked massive waves of pseudoscience and misinformation. I read and watched many interviews with people who worked in public health, infectious diseases, or emergency medicine talking about the importance of staying vaccinated, wearing masks in public, and taking any respiratory symptoms seriously. While they were speaking from years of specialized knowledge and experience, they used simpler language that effectively conveyed their key points to a wide range of people, offering sound (and much-needed) counterpoints to the conspiracy theories and quack medicine running rampant at the time.

The work of those medical professionals and interviewers is just one example of science communication (sometimes abbreviated as "scicomm"). Scicomm can be as varied as a journalist writing about a newly released scientific paper, a park ranger leading an interpretive tour, or a field scientist training volunteers to aid in a survey. We have an expert who is conveying information to laypeople and actively inviting them to connect with the subject matter at hand. Often there is an element of personal impact, demonstrating how the subject matter and the individual affect each other.

Imagine a citizen science project where volunteers are asked to survey an area of land for a certain invasive species. The initial training shows them what to look for and makes them aware of the species' negative impact. The survey itself demonstrates how widespread the problem may be and the relative difficulty of identifying the culprit. The volunteers may then feel compelled to stay engaged, whether through following news stories, sharing information with others, or even volunteering for eradication programs. They now feel that they are a part of something bigger rather than being a simple bystander to science.

Citizen science can also be a great learning opportunity. Many projects require us to hone our nature identification skills within particular taxa or ecosystems. If I am surveying salmon returning to a stream and multiple species may be running at the same time, I need to be able to tell which species I'm looking at so I can keep an accurate count. Positive identification is also incredibly crucial in the case of habitat restoration programs; when I was leading volunteer invasive-plant removal crews for the Friends of Willapa National Wildlife Refuge, I always pointed out how to tell the difference between our target Scotch broom and the native red huckleberry (*Vaccinium parvifolium*), so volunteers wouldn't yank up the wrong plant!

In both cases—surveying and habitat restoration—you won't just be tossed out in the field to fend for yourself. Unless the project specifically calls for people who already have experience, you'll get some training relevant to the task at hand. This might consist of a brief online class, a handout detailing how to identify key species, in-person guidance from a project leader before you head out in the field, or some combination thereof. There will generally be someone around you can ask for help if you aren't sure what species you've found, too. The people who run citizen science projects are used to working with a wide variety of volunteers, so don't let a lack of experience keep you from joining up. Chances are you'll be able to learn as you go and get some great identification practice to boot.

Finding Citizen Science Opportunities

I want to preface this section by pointing out that this is yet another area where options are not equally distributed around the globe. As a lifelong resident of the United States and currently living in the nature-friendly Pacific Northwest, I have access to a wide range of citizen science projects in person. Many of the resources I mention here are specific to the United States or have more projects based here than anywhere else. That said, there are some online options that anyone with an internet connection can participate in, so I cover those as well.

First, identify what citizen science opportunities may be available to you. Websites like CitizenScience.gov, SciStarter.org,

Anecdata.org, and CitSci.org list current projects seeking volunteers; while some are based in the United States, international opportunities exist as well. The Smithsonian Institution lists some global opportunities at si.edu/volunteer/citizenscience as well, from tracking mammals with camera traps to tracking climate change using ginkgo trees. And Zooniverse.org exclusively features online projects open to anyone who wishes to participate.

The data collected by some nature identification apps may be used by professional scientists in their research. eBird is by far the most popular app used by birders, who log lists of species they see out in the field. I've already extolled the virtues of iNaturalist to you; needless to say, the hundreds of millions of observations of species across several kingdoms make a formidable database indeed.

Next, consider what sorts of projects you might like to participate in. The most common activities are on-site surveys, which may focus on individual organisms, numbers of species, their habitats, and habitat restoration, which commonly involve invasive species removal and planting native flora. Some projects are one-off events where the public is invited to participate on a given day or weekend; others may be ongoing, with volunteers either scheduling out blocks of time or dropping in whenever they're able. Either way, you're getting to know the target species in more detail, and you'll likely learn about some of the other living beings that inhabit the ecosystem where you're working, both excellent uses of your nature ID skills.

If you can't do much outdoor activity or don't have any in-person options where you are, how about virtually identifying species you see as you sift through an online database of photos? This helps professional scientists make use of massive amounts of material they wouldn't be able to classify by themselves. Generally, there aren't any minimum time obligations; you can log in any time and participate for however long you like.

These are just a few of your options; more citizen science projects and websites are listed on page 257.

Being a Science Communicator

Once you feel more confident in your identification skills, you may consider finding ways to help others learn more about the nature around them. This can be as simple as informal walks with family and friends or pointing out species you know as you go along. Some parks and other public lands welcome volunteers to assist with or lead tours, classes, and other nature interpretation activities. If you aren't keen on being around people but have decent writing skills, check in with local nature-based nonprofit organizations to see if they need any pieces for their newsletters or social media. (Remember those species reports I recommended in the last chapter? Those can be easily turned into articles!)

Start by contacting local parks, zoos, and other private and public facilities to see if they have or know of any volunteer opportunities. Ask local nature-based nonprofits if they are interested in having someone volunteer as an ecological educator or interpreter. Websites like VolunteerMatch.org and Volunteer.gov are available for those in the United States to search for volunteer activities in their area, and there may be similar websites where you are.

If nothing is in your area, consider going solo! You can start a club or an informal group for those interested in local nature and organize a few walks or other activities. iNaturalist bioblitzes are especially fun, as they're a way to introduce participants to nature identification and to highlight local species. These free events invite people to identify as many species they can in a specific region for a predetermined period of time, and any iNaturalist user can organize one. The more people you have in your group, the more eyes there are to notice cool flora, fauna, fungi, and more!

Finally, for those of you in the United States, I highly recommend seeing if your state has a Master Naturalist program. While the precise curriculum may vary from state to state, generally these programs are aimed at teaching people pertinent information about the ecology of the state, its geological history, its various wild inhabitants, and so forth. For example, I am a certified Oregon Naturalist (the program was formerly known as Oregon Master Naturalists), and the backbone of the program is structured around the Level III ecoregions of Oregon. Naturalists-in-training get to learn about what makes each ecoregion unique, and their fieldwork involves in-depth study of one of them.

Master Naturalist programs are generally designed for people with busy lives; online, self-paced coursework is common, and field courses are usually scheduled for weekends. Both naturalists-in-training and certified Master Naturalists may be required to complete a certain number of volunteer hours every year, and once certified, you may also need to complete annual continuing education hours. I have to log at least forty volunteer hours and eight continuing education hours each year, but the program is flexible in how these may be carried out, so long as they are relevant to my training and work as a Master Naturalist.

As someone who does not have that formal-degree background in the natural sciences, getting my Master Naturalist certification gave me more confidence in my own decades of experience and autodidactic learning. While I certainly don't have the depth and breadth of knowledge of someone with thirty years of fieldwork experience or a PhD, the training I received helped me to contextualize not just the species I had spent many years systematically identifying, but also the habitats and ecosystems of which they were a part. And the emphasis on public-facing volunteering and interpretation offered structure that helped me solidify my career as a nature educator, writer, and tour guide. I already had the skills and knowledge; I just needed a little extra encouragement to get out there and make something happen!

You are not, of course, required to use your nature identification skills for anything other than your own edification. However, should you decide to make more use of what you've learned, I hope this chapter has given you some ideas on what your next steps might be.

leafcutter ants (*Atta* sp.)

The Main Takeaways from This Book

By now, I hope you've had a chance to read over most, if not all, of this book (I know how some of you like to jump around—I can relate!). Each person who reads it is going to get something different out of it; you might find that certain parts are especially useful, or you may realize that the whole thing is exactly what you needed all along. I'd like to reemphasize a few important points:

- Nature identification is a process, not an event. It's going to take time, but you'll get much more reliable identifications as a result.

- Use as many resources as you can. The more sources that agree with you, the more likely your identification is correct.

- It's okay to say, "I don't know." Do your best with what you've got, and if you get stuck in the pursuit of one identification, set it aside for another day.

- "Oops" needs to be a part of any everyday naturalist's vocabulary. If you got an identification wrong and then figured it out later, that's okay! Be forgiving of yourself and remember that errors are just good learning experiences.

- The more you practice, the better you get. The only way to get better at something is to keep doing it again and again. But over time, you'll notice yourself improving, which is a reward in and of itself.

- And, once more because it always bears saying: Never use an app as your only tool for identification. Multiply that by a thousand if you're identifying things you intend to eat. Nature identification is a skill to be honed and practiced over a lifetime and can't be replaced with a few algorithms.

I'm always happy to hear from my readership. If you have questions about anything in this book, need more resources, or just want to share something neat you identified, you can drop me a line at rebeccathenaturalist@gmail.com.

ANIMAL IDENTIFICATION TEMPLATE

Species Observed:

What, When, and Where?

DATE: **LOCATION:**

NUMBER OBSERVED: **LIFE STAGE:**

SEX: **ALIVE OR DEAD?**

HABITAT (be as detailed as possible):

Observed Behaviors

MOVEMENT: **SOUNDS:**

OTHER:

What's It Look Like?

VERTEBRATE OR INVERTEBRATE?

SIZE/PROPORTIONS:

OVERALL SHAPE/DETAIL SHAPES:

COLORS/PATTERNS:

TYPE OF OUTER COVERING (feathers, scales, exoskeleton, etc.):

VISUAL TEXTURE:

TRACKS, SCAT, NESTS, ETC.:

PLANT IDENTIFICATION TEMPLATE

Species Observed:

What, When, and Where?

DATE: **LOCATION:**

NUMBER OBSERVED: **ALIVE OR DEAD?**

VASCULAR OR NONVASCULAR?   **SEEDS OR SPORES?**

ANGIOSPERM OR GYMNOSPERM?   **HABITAT/LIGHT:**

Leaf Details

SHAPE/ARRANGEMENT:

MARGINS:

VEINS:

What's It Look Like?

SIZE/PROPORTIONS:

OVERALL SHAPE/DETAIL SHAPES:

COLORS/PATTERNS:

VISUAL TEXTURE:

TACTILE TEXTURE:

LIFE CYCLE:

FUNGUS IDENTIFICATION TEMPLATE

Species Observed:

What, When, and Where?

DATE: **LOCATION:**

NUMBER OBSERVED: **GROWTH PATTERN (alone, in a group):**

SUBSTRATE (soil, decaying wood, etc.; include species if possible):

HABITAT:

POSSIBLE MYCORRHIZAL PARTNERS:

Other Details

SMELL/TASTE:

BRUISING Y/N (if yes, color?):

DELIQUESCENCE Y/N:

What's It Look Like?

SIZE:

OVERALL SHAPE:

HYMENOPHORE TYPE:

COLORS/PATTERNS:

VISUAL TEXTURE:

TACTILE TEXTURE:

HOLLOW OR SOLID INSIDE:

SPORE PRINT COLOR:

GLOSSARY

achene: A type of dry, simple fruit produced by some *angiosperms,* generally with a single seed.

altricial: Any young animal born helpless and entirely dependent on its parents; nidicolous is a synonym; compare with *precocial.*

angiosperm: A plant that reproduces using flowers and that protects its seeds within ovaries; Magnoliophyta is a defunct synonym; compare with *gymnosperm.*

annulus: A ring of tissue around the stipe of some mushrooms; it is the remnant of a membrane that protected the *hymenophore* while the mushroom was developing; compare with *volva.*

apothecium: A concave (bowl-shaped) *fruiting body* of a fungus or *lichen* containing *spore*-producing structures.

ascomycete: A fungus whose microscopic reproductive structures are sac-shaped; compare with *basidiomycete* and *fungi imperfecti.*

basidiomycete: A fungus whose microscopic reproductive structures are club-shaped; compare with *ascomycete* and *fungi imperfecti.*

binomial nomenclature: The scientific name of a given organism, consisting of its genus and species.

bruising: A discoloration of a mushroom caused by damaging its flesh, and usually appearing within an hour of the damage; staining is a synonym.

bryophyte: A term for *nonvascular* plants that may refer only to the mosses (Bryophyta), or collectively the mosses, liverworts, and hornworts.

cap: The top portion of a mushroom that is supported by the stipe and that itself contains the *hymenophore,* typically on its underside; pileus is a synonym.

citizen science: Science in which nonprofessionals from the general public assist professional scientists in data collection and other research activities, generally as volunteers.

cloaca: An orifice found in amphibians, reptiles, birds, and a few mammals through which the animals urinate, defecate, and lay eggs or give birth to young.

convergent evolution: When unrelated species evolve similar traits independently; these species may live at the same time or at different points in the history of the planet.

cotyledon: An embryonic leaf that is generally the first to appear when a seed germinates; see *monocot* and *dicot.*

deciduous: Describes plants that lose their leaves seasonally.

decomposer: An organism that exudes digestive enzymes onto decaying matter and then absorbs nutrients from it; compare with *detritivore.*

decurrent: Describes the *hymenophore* of a fungus when it extends from the cap partway down the stipe.

detritivore: An organism that consumes entire fragments of decaying matter and digests it internally; compare with *decomposer*.

deuterostome: Any of a number of animals that, as embryos, develop the orifice that will become the anus before the one that becomes the mouth.

dichotomous key: An identification tool that leads the user through a series of choices concerning the traits of an organism; the decision made at each one determines where in the key to go next and ultimately should lead to the identification of the organism at hand.

dicot: Any flowering plant whose first sprout produces two *cotyledons;* eudicot is a synonym.

diploid: A cell that contains two copies of its chromosomes, generally one from each parent organism; compare with *haploid*.

drupe: A type of fruit consisting of a fleshy exterior surrounding a single seed wrapped in a hard shell; peaches are an example of a drupe, while an aggregate fruit like a raspberry is made of multiple drupes (or drupelets).

DYC: "Damned yellow composite"; any of a number of similar flowers with yellow petals in the family Asteraceae that can be difficult to tell apart.

eukaryote: Any organism whose cells have nuclei wrapped in membranes; compare with *prokaryote*.

field guide: A book containing pictures and common and scientific names; it describes unique characteristics of a specific group of organisms and is used to help people identify the species described therein.

frass: The solid waste material of insects.

fruiting (or fruit) body: The reproductive portion of a fungus or *lichen;* mushroom is a synonym specifically for the fruiting bodies of many *basidiomycete* and *ascomycete* fungi.

fungi imperfecti: A fungus that is neither an *ascomycete* nor a *basidiomycete*.

gametophyte: A life stage of plants and algae in which they grow reproductive structures that then reproduce with *haploid* sex cells; compare with *sporophyte*.

gemmae: Asexual reproductive structures of some plants that reproduce through fragmentation or budding; primarily seen in *nonvascular* plants, although a few *vascular* plants also produce gemmae.

gymnosperm: A nonflowering *vascular* plant that does not encase its seeds in an ovary and that may be contained in a cone or other similar structure; compare with *angiosperm*.

haploid: A cell that contains one copy of its chromosomes, commonly (although not universally) in preparation for combining with another haploid cell; compare with *diploid*.

herp: Short for herpetofauna, it refers collectively to reptiles and amphibians; the study thereof is known as herpetology.

holobiont: A symbiotic system centered on a primary host with several other species, creating a miniature ecosystem; *lichens* are an example.

hymenophore: The spore-producing structure of a mushroom or other fungal *fruiting body,* which may take the form of gills, pores, or other shapes, and that contains the hymenium, a special tissue from which grow *spore*-producing structures.

hyphae: Microscopic filaments that make up the *mycelium* of a fungus.

invertebrate: Any animal that does not have a *notochord* or that may otherwise be defined as "not a *vertebrate*"; compare with *vertebrate.*

isidia: Small fragments of a *lichen* that may break off and grow into new lichens as a form of asexual reproduction; compare with *soredia.*

LBJ: "Little brown job" (alternately, LBB, "little brown bird"); any of a number of small, nondescript brown songbirds that can be difficult to tell apart.

LBM: "Little brown mushroom"; any of a number of small, plain brown mushrooms that can be difficult to tell apart.

lichen: A symbiotic being composed of a primary fungus species, a *photobiont,* and a yeast fungus; other undescribed species may be involved, and lichens might be an example of a *holobiont.*

merosity: The typical number of parts a plant's flower may have; generally, all parts of a flower, such as petals, sepals, stamens, and so on, occur in the same number or multiples of that number; in some cases, other parts such as leaves may also display merosity.

metamorphosis: A series of life stages in some *invertebrate* animals in which each stage is distinct and may vary widely from previous or following stages; hemimetabolic metamorphic stages all resemble smaller versions of the adult, while the younger stages in holometabolic metamorphosis may be drastically different than the adult.

microphyll: A type of simple leaf that has only one central vein.

monocot: Any flowering plant whose first sprout produces one *cotyledon;* compare with *dicot.*

mycelium: The primary, nonreproductive portion of a fungus made of tiny filaments called *hyphae,* which grow throughout whatever substrate the fungus prefers; the *thallus* of lichens is made of mycelium.

mycobiont: The primary fungal partner or host of a *lichen;* compare with *photobiont.*

mycorrhizal fungi: A species of fungus that has a mutualistic relationship with a plant; the *mycelium* of the fungus intertwines with and may even penetrate the plant's roots to facilitate the sharing of resources.

nonvascular plant: A plant that does not have vascular tissues like xylem or phloem, but instead possesses a simpler system for transporting water and nutrients.

notochord: A flexible rod-like structure found in all vertebrates at some stage of development, although in most it is replaced by the vertebral column and exists only postnatally as part of the intervertebral discs.

perithecium: A flask-shaped structure that releases a lichen's *spores.*

phloem: *Vascular plant* tissue that transports photosynthesized compounds from the leaves to the rest of the plant, including storage in the roots; compare with *xylem.*

photobiont: The photosynthesizing partner of a *lichen,* usually an algae or cyanobacteria and sometimes both; compare with *mycobiont.*

phyllid: A structure on a *nonvascular plant* that is similar to a leaf but simpler, and generally only a single cell thick.

phylogenetics: The study of how various organisms are related to one another throughout the history of their evolution.

precocial: Any young animal born with varying levels of independence, from being able to walk and follow a parent from birth to being entirely capable of caring for itself; nidifugous is a synonym; compare with *altricial.*

prokaryote: Any organism whose cells do not have nuclei or other organelles wrapped in membranes; compare with *eukaryote.*

protist: Any *eukaryote* that is not an animal, plant, or fungus.

protostome: Includes most animals that are not *deuterostomes;* originally, all protostomes were assumed to form their mouth first and then their anus, but more recent research demonstrates much more variation among this group.

rhizoid: A simple root-like structure that helps *nonvascular plants* anchor themselves to their substrate.

rhizome: A type of plant stem that grows laterally underground and produces entirely new plants as a form of asexual reproduction.

scat: Another term for an animal's fecal matter, particularly that of terrestrial *vertebrates.*

seed: The *diploid gametophyte* of a plant, the product of sexual reproduction combining two *haploid* sex cells; the embryo of the plant resides within a protective shell (testa) along with sufficient food for development and germination; compare with *spore.*

senescence: In plants, the seasonal aging and loss of leaves and other temporary structures.

sexual dimorphism: The tendency of individuals of various sexes to have markedly different characteristics from one another; most often seen in female and male animals of some species.

soralia: A cluster of *soredia.*

soredia: Masses of cells from both the fungal and *photobiont* members of a *lichen* that may break off and grow into new lichens as a form of asexual reproduction; compare with *isidia.*

sori: A mass of *sporangia,* such as those found on the underside of fern leaves.

sp.: the abbreviated form of "species"; denotes the highest taxon of organisms that can still typically reproduce with each other and share the majority of genetic data; compare with *spp.*

sporangium (plural *sporangia*): The organ in plants, fungi, and other organisms that produces *spores.*

spore: A *haploid* cell produced by plants, fungi, algae, and slime molds and other protists; this may be part of a larger sexual reproductive cycle; compare with *seed.*

sporophyte: A life stage of plants and algae in which they asexually reproduce with *diploid* cells (*spores*); compare with *gametophyte.*

ssp.: The abbreviated form of *subspecies.*

stipe: The stem of a mushroom that supports the *cap.*

strobili: Stalk-like structures on some nonflowering plants that produce *spores* from clusters of *sporangia.*

subspecies: A smaller division of organisms within a given species that display consistent unique traits compared to other members of the same species.

taxon (plural *taxa*): A particular division of living organisms. The seven primary taxa are, from the most general to the most specific: kingdom, phylum, class, order, family, genus, *species.*

taxonomy (biology): The science of classification, particularly of various living organisms, based on shared characteristics.

tetrapod: Any *vertebrate* with four limbs or descended from such (as in snakes, for example); includes all amphibians, reptiles, birds, and mammals, both extant and extinct.

thallus: The nonreproductive portion of a fungus, *lichen,* slime mold, algae, and some liverworts.

var.: The abbreviated form of *variety.*

variety: A smaller division of organisms within a given *subspecies* that display consistent unique traits compared to other members of the same subspecies.

vascular plant: A plant with a vascular system consisting of a *xylem* and *phloem;* compare with *nonvascular plants.*

vertebrate: An animal with a *notochord,* although a few do not have vertebrae in spite of their phylum name, Vertebrata; Craniata ("animals with skulls") has been proposed as a more accurate term; compare with *invertebrate.*

volva: A cup of tissue at the base of some mushrooms, particularly within the family Amanitaceae; it is the remnant of a membrane that protected the mushroom during development; compare with *annulus.*

xylem: *Vascular plant* tissue that transports water and nutrients from the roots to the leaves, stems, and other extremities; compare with *phloem.*

RECOMMENDED RESOURCES

These are just a few of my favorite identification resources. Some of them are canted toward North America; others have more widespread applicability. They're best used in conjunction with more region-specific materials like field guides, of which there are far too many to list here. Many of these have been mentioned elsewhere in the text, but I wanted to have them all in one convenient place to make it easier to look them up again.

Books

Kaufman Field Guide to Advanced Birding: Understanding What You See and Hear by Kenn Kaufman (Houghton Mifflin Harcourt, 2011): This is the book I describe to my birdwatching students as "Birding 201." Kaufman does an excellent job of explaining how to identify various bird taxa, including tricky ones like gulls and sparrows. The book also discusses variations in plumage due to factors like molting and age. While the species are North American, there is plenty of material that may be useful to folks outside that continent. Please note that this is not a field guide of individual species, but a book of patterns and traits to look for, just as I detailed in the How to Identify Animals chapter.

Botany in a Day: The Patterns Method of Plant Identification by Thomas J. Elpel (HOPS Press, 2013): While the title is a little hyperbolic—you won't learn everything in the book in a day—this is an excellent resource for learning how to identify plants down to the family level by looking at traits like the structure of their flowers, leaf shape and arrangement, and stem shape. While the genera that Elpel uses as examples in each family are drawn from North America, they are often found elsewhere in the world and the patterns he describes still generally hold true.

Plant Identification Terminology: An Illustrated Glossary by James G. Harris and Melinda Woolf Harris (Spring Lake, 2021): This exceedingly thorough book clearly defines pretty much every plant-related term you're likely to encounter in your nature identification endeavors. The illustrations show what to look for on the plant itself, making this an invaluable tool, especially for those wishing to work with dichotomous keys.

How to Identify Plants by Harold D. Harrington and illustrated by L. W. Durrell (Swallow Press, 1985): This is a good companion to the Harrises' work above as it also details various parts of plant anatomy with illustrations. It also contains some good tips for collecting and preserving plant specimens for further study, should you ever have reason to do so.

A Mushroom Word Guide: Etymology, Pronunciation, and Meanings of Over 1,500 Words by Robert M. Hallock (self-published, 2019): If you need a glossary of fungal terminology, this is a great option. I like that the etymology is included for many of the words; it often becomes a clear "Oh, that's why we use this word for that part of a mushroom!" And for those of us who have read but never heard countless words (which we invariably end up mispronouncing in front of an audience), the pronunciations are quite welcome. While sometimes tougher to find, *How to Identify Mushrooms to Genus I: Macroscopic Features* by David L. Largent and Daniel E. Stuntz (Mad River Press, 1986) makes a great companion book for this one. It is a dichotomous key, but it has illustrations, something Hallock's work does not have. And even if you don't live in the Pacific Northwest, *Mushrooms of Cascadia: An Illustrated Key to the Fungi of the Pacific Northwest* by Michael Beug (Ten Speed Press, 2024) has terrific photos detailing the various minute details of fungal anatomy, so you know what you're supposed to be looking for.

A Field Guide to Bacteria by Betsey Dexter Dyer (Comstock, 2003): If you're interested in seeing life through a microscope, this is an invaluable guide. Dyer details how to see and identify the bacteria growing on old food, living in soil and water, and even infecting plants. The appendices include useful information on culturing bacteria, using microscopes, and more.

Websites

AllAboutBirds.org: This is my go-to website for North American birds. Each species profile includes identification information, pictures, videos, sound files, and comparisons with similar species.

Academy.AllAboutBirds.org/bird-academys-a-to-z-glossary-of-bird-terms: Run into a bird-related term you don't recognize? Chances are it's in this online glossary that goes from "abdominal air sacs" to "zygodactyl feet."

HerpMapper.org: This is an excellent site for identifying reptiles and amphibians worldwide—if you know the species. You can search the site for all records of that species that users have uploaded over the years and compare them to your mystery herp. Create an account, and you can upload pictures or sound files of reptiles and amphibians for identification or confirmation by other users and inclusion in the database.

BugGuide.net: This is not the easiest site to use if you have a completely unknown arthropod in the United States or Canada and you just want to find pictures of similar ones. Click on "Guide" and then choose the appropriate subphylum, class, order, and so forth, using the sample photos of each taxa to help you make your decision. However, if you are pretty sure you know what genus or species you have, you can search for it on the site and get lots of pictures for comparison. You can also create an account and submit images of arthropods for identification help, but if the photos are not of good quality, they will end up removed from the site after a month.

FloraNorthAmerica.org: This is another site that works best if you already have some idea of what you have and it exists in the United States or Canada. The search function works best at the genus or species level if you want to double-check your identification; beyond that, there may be too many pictures to sort through, especially for large families like the Asteraceae. You can also browse the site by taxonomic hierarchy, but this can be overwhelming if you don't know your species' class, order, and so on.

WildflowerSearch.org: I like this tool better for North American plant ID if you have no clue what you've got besides "wildflower," "fern," "conifer tree," and so on. The left sidebar allows you to enter the coordinates and elevation of the plant you're identifying, and then narrow down options by traits like what kind of plant it is (seaweed and lichens are included, too), flower color and shape, and leaf attachment. To the right will be photos of plants matching the parameters you specified.

MushroomExpert.com: Trying to figure out a North American mushroom species? Not only does this site profile more than 1,200 species, but there's also a pretty user-friendly key, too. You may need to look at spores under a microscope to get their shape, though—something that is unavoidable with some species' identification.

MushroomObserver.org: Create an account here, and then you can upload observations of fungi that other users can then help identify. Or if you already know what it is, include the species and give detailed information about why you identified it as such. Once you're more confident, you can help others with their observations, too.

MicrobeHunter.com: If you are at all interested in exploring tiny life-forms, this website is invaluable. There's plenty of information on choosing and using a microscope, preparing slides, seeing specific organisms like tardigrades, taking photos of what you find, and more. It does not contain an identification guide, but the forums are active and may be a good place to ask for help.

Online Communities

These are all groups that are free to join on Facebook or Reddit. You can find the Facebook groups by searching for their names in the search bar on the site; for Reddit groups, type Reddit.com and then the subreddit name, which always starts with /r/ (for example, Reddit.com/r/WhatsthisBird). Some of the larger, more general Facebook groups may be so overloaded with posts that there's a good chance yours may get buried by the algorithms, so I've listed some here that are big but not too big. I recommend looking for groups that are more specific to your area whenever possible. Once you have some practice at identification, consider helping others identify the organisms they post.

Birds

What's This Bird?—American Birding Association (ABA) (Facebook)

North America–centric bird ID group; it's meant to be beginner-friendly.

People Helping Others ID Birds (Facebook)

/r/WhatsThisBird (Reddit)

Both these sites are open to anyone around the world to post pictures of birds they're having trouble identifying, although they both tend to have many North American users. Make sure you include your location when posting pictures or videos.

Reptiles and Amphibians

Reptile and Amphibian Identification (Facebook)

/r/herpetology (Reddit)

If amphibians and reptiles are your thing, here's where to get them identified.

Arthropods

The Entomology Public Group (Facebook)

/r/WhatsThisBug (Reddit)

These two groups are available for arthropod identification around the world; you can also ask questions or just post pictures of neat arthropods you've already found and identified.

Tracking and Scatology

Animal Tracking and Scatology (Facebook)

Scats, Tracks, and Animal Identification (Facebook)

/r/AnimalTracks (Reddit)

All three groups are good resources for getting help identifying animal tracks and scat, although the Facebook groups are more active as of this writing.

Plants

People Helping Others ID Plants (Facebook)

/r/WhatsThisPlant (Reddit)

Both these groups may be used to identify both native and nonnative plants around the world.

Trees

Identify That Tree (Facebook)

A worldwide group for helping you figure out what tree you have based on pictures of leaves, flowers, or the entire tree.

Mushrooms

The Mushroom Identification Group (Facebook)

/r/Mushrooms (Reddit)

Both are great places to get help figuring out those fiddly fungi, plus their About sections include further website links and other valuable resources about mushrooms.

Poisons Help

Poisons Help; Emergency Identification for Mushrooms & Plants (Facebook)

This is a useful global resource if you, another person, or a pet eats some plant or mushroom that makes them sick or something else you are concerned about (or call Poison Control at 1-800-222-1222 for emergencies). Scrolling through the posts is also an educational experience, particularly on what *not* to eat and how dangerous (or not) various species may be.

Apps

iNaturalist: In case you haven't figure it out by now, this is my all-time favorite nature identification app. Animals, plants, fungi, slime molds—if it's alive and you can take a picture or a sound file of it, you can upload it to iNaturalist. In addition to initial algorithmic ID recommendations, you can also get suggestions, verifications, and corrections from other iNaturalist users, which include many experienced naturalists in a variety of fields. While you should never use an app as your only ID tool, this one is the best of the best.

eBird: While you can also upload pictures and sounds of birds to this app, it is a lot less efficient than iNaturalist. eBird primarily allows you to make checklists of birds you identify at a given place and time. You then have to go back into that checklist by logging in on your browser and manually uploading any images or sound files you took during that trip to the appropriate species in your list. However, if you are a bird fan, eBird is a definite boon as you can see other people's checklists—and look for those rare species sightings!

Merlin Bird ID: While iNaturalist does identify animals by sound, this is my absolute favorite for North American bird calls because the app listens to them live and tells you what they are as they vocalize. You still want to double-check Merlin's IDs with another tool like AllAboutBirds.org sound files, but I've had good experiences with this app. And it saves your recordings so you can go back and listen to them later.

ACKNOWLEDGMENTS

My thanks to my agent, Jane Dystel of Dystel, Goderich & Bourret, LLC, for partnering with me to find an excellent home for this book, to Kristin Hugo for introducing us in the first place, and to Julie Bennett, my editor at Ten Speed Press, for helping me fine-tune the manuscript into something I can truly be proud of. Mi Ae Lipe showed exactly why copy editors are so valuable, and caught a terrifying number of typos, oopses, and errors I missed in the manuscript, while Lisa Brousseau and Mindy Fichter further refined my work with thorough proofreading. I very much appreciate the work that Francesca Truman put into designing this book, and the incredible illustrations by Ricardo Macía Lalinde that bring such life to these pages. *The Everyday Naturalist* also couldn't have happened without the expert input from production editor Sohayla Farman, production manager Philip Leung, and art director Betsy Stromberg. Getting the word out about it before its launch was much easier with help from marketer Monica Stanton and publicist Kristin Casemore. And thank you to Lissa Brewer, who provided me with some amazing author photos on an excursion through old-growth cedar forests.

Much gratitude goes to Syren Nagakyrie of Disabled Hikers for helping me find more resources for disabled naturalists and for being an inspiration as an author and educator. My unending appreciation to the Oregon Naturalist program for helping me gain a deeper ecological understanding of this area, and to all the community colleges, libraries, and other entities who have hosted my online and in-person classes over the years. I also want to thank my parents, Marianne and Dave Lexa, for instilling in me a love of writing and being my first editorial team from first grade onward. Finally, I cannot say enough good about the immense, irreplaceable support my partner, Stephen, has been for well over a decade now; thanks for being my biggest cheerleader through it all.

INDEX

Note: Page numbers in *italics* reference illustrations.

ABOUT THE AUTHOR

REBECCA LEXA, MA, OMN, WFR, has made it her goal to spend as much time as possible connecting people to the natural world. With a background in ecopsychology, she is additionally certified as an Oregon Naturalist and Wilderness First Responder. An enthusiastic and approachable science communicator, her teaching and speaking engagements in the Pacific Northwest and beyond have reached thousands of everyday naturalists of all ages, and her guided nature tours offer deeper hands-on experience with local ecology. More about Rebecca and her work may be found at RebeccaLexa.com.

© LISSA BREWER

Published in the United States by Ten Speed Press, an imprint of the Crown Publishing Group, a division of Penguin Random House LLC, New York.
tenspeed.com

Ten Speed Press and the Ten Speed Press colophon are registered trademarks of Penguin Random House LLC.

Typefaces: Monotype's Gazpacho, Mostardesign's Strato Pro, and Linotype's Helvetica

Library of Congress Cataloging-in-Publication Data is on file with the publisher.

Trade Paperback ISBN: 978-0-593-83597-5
eBook ISBN: 978-0-593-83598-2

Printed in China

Editor: Julie Bennett | Production editor: Sohayla Farman
Designer: Francesca Truman | Art director: Betsy Stromberg |
    Production designers: Mari Gill and Faith Hague
Production manager: Philip Leung
Copyeditor: Mi Ae Lipe | Proofreaders: Lisa Brousseau and Mindy Fichter |
    Indexer: Jay Kreider
Publicist: Kristin Casemore | Marketer: Monica Stanton

10 9 8 7 6 5 4 3 2 1

First Edition